About th

A writer known for her descriptions of colourful characters and living legends, Shirlee Smith Matheson has published twelve books, both adult nonfiction and young adult fiction. Shirlee has been described as a writer who has true feeling and understanding for her subjects, especially those in her *Flying the Frontiers* series, in which she sets down the adventurous histories of Canadian bush pilots and engineers. Other books by Shirlee include the award-winning *This Was Our Valley*, about the impact of WAC Bennett Dam in northern BC, and *Youngblood of the Peace,* the story of an Oblate priest who served the First Nations people of British Columbia and Alberta. Shirlee is also a respected writer of juvenile and young adult books, with award-winning titles such as *Flying Ghosts, The Gambler's Daughter, Keeper of the Mountains*, and *Fastback Beach.*

Shirlee Smith Matheson has taught evening writing courses at the Alexandra Writers Centre and Alberta College of Art and Design in Calgary, Alberta, and has been Writer in Residence at libraries, schools, and colleges throughout Alberta. She is a charter member of Women in Aviation International, Rocky Mountain High Alberta/NWT Chapter, and is Administrative Officer at the Aero Space Museum of Calgary. Research and reading tours take Shirlee across the country to schools, libraries, and museums, and into the homes and hangars of the aviators whose biographies are featured in her *Flying the Frontiers* series. You can find out more about Shirlee and her work at www.ssmatheson.ca.

To my nephew Harley.   Sept. 9/05.

I send this to you with our Love and
Gods Blessing, as well as loveing, proud
fond memories of my Brother, Your Dad
George. I know you share these!

Shirlee has become a dear friend
since I first contacted her and then met
her some 7 or 8 years ago! We talk

This book is dedicated to those who have experienced
the serious side of flying.
Some returned, some did not, and many remember.

Time, like an ever-rolling stream,
Bears all its sons away;
They fly, forgotten, as a dream
Dies at the opening day. ★

on the phone about once every month
or two.   Take care,
         Love from Uncle Jack.

★ Processional Hymn, Dr. Wm. Croft, 1708,
descant by Dr. Alan Gray, from *The Book of Common Praise*
(being the hymn book of the Anglican Church of Canada,
Toronto, Oxford University Press, revised 1938.)

September 2005

To Harley Greening,

With your father's poem so well expressing the hopes of airmen who find themselves lost, but not alone, I thank you.

Shirlee Smith Matheson

LOST

True Stories of Canadian
Aviation Tragedies

Shirlee Smith Matheson

**FIFTH
HOUSE**

Cover and interior design by Kathy Aldous-Schleindl
Edited by Lesley Reynolds
Copyedited by Kirsten Craven
Proofread by Ann Sullivan
Scans by ABL Imaging

The publisher gratefully acknowledges the support of The Canada Council for the Arts and the Department of Canadian Heritage.

THE CANADA COUNCIL | LE CONSEIL DES ARTS
FOR THE ARTS | DU CANADA
SINCE 1957 | DEPUIS 1957

We acknowledge the financial support of the Government of Canada through the Book Publishing Industry Development Program (BPIDP) for our publishing activities.

Financial assistance was provided to the author by the Alberta Foundation for the Arts.

Alberta Foundation for the Arts

Printed in Canada by Friesens

05 06 07 08 09 / 5 4 3 2 1

First published in the United States in 2006 by
Fitzhenry & Whiteside
121 Harvard Avenue, Suite 2
Allston, MA 02134

Library and Archives Canada Cataloguing in Publication
Matheson, Shirlee Smith
        Lost : true stories of Canadian aviation tragedies / Shirlee
 Smith Matheson.
Includes bibliographical references.
ISBN 1-894856-18-X
        1. Aircraft accidents--Canada--History.   I. Title.
TL553.5.M38 2005   363.12'4'0971   C2005-902363-5

Fifth House Ltd.
A Fitzhenry & Whiteside Company
1511, 1800-4 St. SW
Calgary, Alberta T2S 2S5

1-800-387-9776
www.fitzhenry.ca

CONTENTS

INTRODUCTION

Reverie

I often wonder as I fly
Where go the airmen when they die?
Maybe they don't die at all
But only move to another haul.

Often on days when the sky is bright
And lofty pinnacles grace the height,
I wonder if in the upper air,
My pilot friends are waiting there.

In Heaven they'd not do menial things
But course the sky on special wings.
Ready to help us guys below
Guiding us when big storms blow.

Often when I have judged it wrong
And needed help to get along
I wonder if airmen gone before
Have guided me safely down once more.

George Greening[1]

In researching *Lost: True Stories of Canadian Aviation Tragedies*, I discovered I was mining a very rich field. Considering the number of ghostly tales, unexplained sightings, and mysterious accidents and disappearances connected with the "flying field," it is not surprising that at least four patron saints have been dedicated to overseeing the safety of aviators: Joseph of

Cupertino, Our Lady of Loreto, Therese of Lisieux, and St. Raphael.

This collection of stories covers aviation anomalies that have occurred across the country, from the coasts of Labrador to British Columbia. Some reach beyond our aerial borders to chronicle Canadian participation in solving aviation mysteries, and bringing resolution, no matter how dire, to disappearances and disasters.

What happened to Johnny Bourassa, a former World War II air force hero, en route in the Bellanca from Bathurst to Yellowknife? Where is the crew of the famous "*Stalin's Falcons*," who disappeared near the North Pole on their flight from Moscow to the United States in a huge four-engined bomber? Are there really "haunted" flight paths, or is the high accident rate over the mountains from British Columbia's coast to Alberta just one of the roughest, toughest air spaces in the world? From disappearances of hockey heroes to veteran northern pilots, these stories revisit tales still recounted in transport hangars and offices with wonder, awe, unending questions, and reverence for the lost souls of the aviation fraternity.

The author is grateful to many people around the world who provided information and research documents for this project. Their names are noted in the stories and source lists. Special thanks go to Clark Seaborn, P.Eng., who provided valuable technical editing. Research for this book took me into private homes, hangars, and offices to interview participants or their surviving families, friends, and flying associates. It was not an easy task, as these disasters and disappearances often had deadly results. But the stories are real and true, and they chronicle flights—and flight plans—that have gone terribly wrong.

FLIGHT PLANS FOR FREEDOM

On Friday, December 14, 1979, Ken Leishman was booked for a medivac to fly a Sabourin Lake Airways Ltd. twin-engine Piper Aztec from Red Lake to the First Nations community of Sandy Lake and back to the hospital at Thunder Bay, Ontario. His landing at Sandy Lake was smooth. Jackie Meekis, a thirty-five-year-old community health nurse, helped load her fifty-one-year-old stretcher patient, Eva Harper, who'd broken her hip, and at 5:05 PM they took off. Ken called in to the Thunder Bay air traffic control tower at 8:20 PM, stating he was at five thousand feet and preparing to descend to three thousand feet. Permission was granted. At 8:40 PM the white aircraft with black markings disappeared.

A search party was immediately organized comprised of military, civilian, and Ontario Provincial Police forces. Although the temperatures hit –30°C, it was assumed that if the pilot and his passengers were still alive following a crash, the cold would not be a problem as they had emergency rations, sleeping bags, and blankets. All they lacked was an Emergency Location Transmitter (ELT), which at the time was not required by law.

When few clues emerged to reveal what might have happened to the aircraft, news reporters sought to find out everything they could about the pilot. Ken Leishman had a good aviation record and was on call to Sabourin Lake Airways to undertake some of their emergency medical flights. But Ken's personal past made for much more exciting news—this man was internationally known as the "Flying Bandit"! Stories about the famous escape artist would sell papers all over the world.

No trace of the aircraft was found, and on Christmas Eve the search was called off, with a plan to resume when spring weather melted the snow.

Ken Leishman had disappeared.

SETTING THE STAGE

William Kenneth Leishman was born on July 20, 1931, in Kenville, Manitoba, west of Winnipeg on Highway 2, and grew up in nearby Treherne. By all accounts Ken had a tough childhood. He quit school early and headed to Winnipeg, where he looked up his father, Norman Leishman, whom he hadn't seen in years. Norman helped Ken get a job installing and repairing elevators. Later, when Norman became owner of Mid-West Elevator and Motor Co., he invited Ken to join him.

Ken had grown into a handsome man, six foot two inches tall and slender, prematurely bald, but sporting a rakish Clark Gable moustache. His physical attributes, along with his natty clothes, gift of the gab, and pleasant manners made him instantly likable and a top salesman.

In 1949 Ken met Elva Shields from Somerset, near Treherne. Although he was just eighteen and she twenty, they were married on February 25, 1950. The young couple rented an unfurnished apartment in Winnipeg, but it was rather bare and Ken wanted to make things nice for his bride. One night, while repairing an elevator at Genser's furniture warehouse on Portage Avenue, Ken realized that all of the employees had gone home and he was locked in. He gathered up some furniture and called a delivery truck. "The store has a rush order," he announced. "We need a load delivered right away." He told the truck driver that a westbound semi-trailer would meet them at a downtown Texaco station. Ken offered to accompany the driver to help off-load the furniture, and then stay with the load in case the truck was late. Once the driver had left the scene, Ken borrowed a friend's pickup truck to assist in the transfer of goods, and by that evening his and Elva's apartment looked much brighter. Elva was innocently impressed. Nothing appeared in the papers; no cops came knocking.

When Elva was expecting their first baby, Ken repeated the operation. He hid in Genser's elevator shaft until all of the employees left, then selected lamps, coffee tables, a refrigerator, and a stove. This time he decided to call

Ken and Elva Leishman's wedding picture, February 25, 1950. (Courtesy Leishman Family)

a different transfer company. The truck came to the back door with two drivers, and Ken told them what to load. At that point they revealed themselves as policemen, called by a suspicious dispatcher. Ken was arrested and taken to Rupert Street station, photographed, fingerprinted, and charged with theft. He wept. Elva also wept when she learned how Ken had procured the furniture, and it was all taken away. Ken pleaded guilty to two theft charges and received a nine-month sentence in Headingly Provincial Jail. He served four months and was released on parole in time to welcome the birth of his daughter, Lee Anne Mae, on September 22, 1950.

FLYING HIGH: THE 1950s

Following several setbacks where his criminal record blocked job opportunities or any advancements, Ken's mechanical and agricultural background gained him a job at Machine Industries Ltd. selling farm equipment. Sales were good, but it took forever to travel the back roads to the various farms. Why not fly?

Ken asked about flying lessons at the Winnipeg Flying Club—in a 65-horsepower Taylorcraft—and at the same time spotted a red Aeronca Chief aircraft for sale for one thousand dollars. He signed up for the required thirty hours of flying lessons at Graffo Flying School and bought the Aeronca. After five hours of lessons and no licence as yet, he loaded the aircraft with machine parts and took off, barely clearing the airport fence and surrounding buildings. When he landed in a farmer's field near Yorkton, Saskatchewan, he was an instant hit. By the end of summer 1953, he'd flown over most of Saskatchewan and Alberta, servicing and selling farm equipment. However, on his return to Winnipeg that September, he was charged with flying without a pilot's licence and received a one-hundred-dollar fine and a two-year suspended sentence.

When Machine Industries was taken over by the Bristol Aircraft

Company, Ken became lead hand in their electrical department, working on Mitchell bombers, Beechcraft Expediters, and Mustang fighters. The hangar was located in Carberry, 135 miles from Winnipeg, so the Leishmans moved back to Treherne, thirty miles from Carberry, and Ken, now duly licensed, flew to work. One day Ken was summoned to the general manager's office, where he was greeted by two RCMP officers who read out a list of his previous convictions. "This plant is engaged in classified work for the government and you are considered a security risk," the manager said. "I'm sorry to say you are no longer an employee of this firm."

So Ken was forced to go back to work for his father. With the elevator controls supplied by Anderson Elevator Co. in New York, and the cylinders and pumps by Lesco Mfg. in Dayton, Ohio, Ken was flying again, this time to destinations south of the border. But the usual problems surfaced with his dad's company—no money.

He next landed a job peddling Queen Anne stainless steel cookware on commission, door-to-door or wherever the airplane could take him, and quickly became the company's top salesman in Western Canada. In the meantime, the Leishman family had increased to four children with the births of Ronald William Stewart (December 16, 1951), Dale Elton (June 18, 1953), and Wade Kenneth (March 23, 1955). A promotion with Queen Anne led Ken and his family to Saskatoon, Saskatchewan, where son Blair Allan was born on April 3, 1956.

When the job with Queen Anne collapsed, the family returned to Winnipeg, where Ken worked for a cookware competitor, Rena-Ware Co. The future again looked promising. Then one morning the managers paid him a personal visit. "Because of the company's high standards, we cannot keep a man on our staff who has a prison record." Ken was fired again.

Although he was forced to sell the Aeronca to pay his debts, Ken had by now logged the required number of training and flying hours to qualify for a commercial pilot's licence. He decided to apply for a job with an airline. When told that airlines required their pilots to have at least grade eleven equivalence, he took a test with the department of education and received the required certificate. His application and subsequent interview with Canadian Pacific Airlines in Vancouver did not even get him on the short list. His prison record interfered once again and he was back to selling cookware, this time with Aristocrat. When sales increased, he bought a three-place Stinson aircraft and proudly painted "Aristocrat" along both sides. He was again flying and selling. Sales progressed and Ken was promoted to branch manager, but without forewarning, the company terminated its Winnipeg operation.

Next, Ken sold twin-engined executive Apache aircraft for a Piper Aircraft dealer, and bought a four-place Stinson Station Wagon for himself. By 1957 he owned a house in River Heights on the outskirts of Winnipeg and a two-year-old Cadillac. He had a thriving family and was an active community volunteer—even organizing Winnipeg's float in the Grey Cup parade in Toronto, towed by his shiny Caddy. Ken had also become well known in the aviation community and when a civil pilots' Volunteer Air Patrol was formed, patterned after the American Civil Air Patrol, he was appointed as district, then national director.

He could now activate a lifelong dream: the construction of Leishman's Fly-In Fishing Lodge. He scouted a location and arranged a ninety-nine-year lease on the shores of the remote and lovely Moar Lake, two hundred miles north of Winnipeg, then made dozens of trips to bring in materials and supplies for the lodge. When his savings were spent, he approached his bank for a loan, but was turned down flat, forcing him to deal with a high-interest finance company. Now Christmas was coming and there were no airplane sales. He couldn't even sell his own Stinson, and the finance companies turned him down for further advances. Ken needed money, and fast.

VISITS TO THE BANK

In desperation over losing everything he'd worked so hard to achieve, Ken made a plan. First he drove to Fargo, North Dakota, and purchased a .22 caliber pistol. Then he borrowed two hundred dollars from a friend and booked a return ticket to Toronto on a Trans Canada Airlines (TCA) flight leaving Winnipeg on December 16, 1957.

He took a day to case the Toronto streets and selected a busy Toronto Dominion bank on the corner of Yonge and Albert. Dressed in a thick wool overcoat, silk scarf, and stylish homburg, Ken introduced himself at the bank as Mr. Gair, a lawyer from Buffalo, and asked to see the manager. He had an interesting proposal to discuss and wished to speak to him privately. Mr. A. J. Lunn welcomed Mr. Gair into his office and went to hang his coat and hat on the stand. He turned around to find himself staring into the barrel of a gun.

"Mr. Lunn, I'm sure you realize that this is a holdup."

With the pistol concealed by his coat draped over his arm, Ken followed the stunned manager out to the cashier's wicket. The requested sum of ten thousand dollars was issued in ten, twenty, fifty, and one hundred dollar bills, bound with an elastic band. Then, Ken and the manager walked outside. "I don't understand you," the manager said, barely able to control his nervous frustration. "You seem to have all the qualities of a business executive. You could do any job, and yet you are doing this. Why?"

"Mr. Lunn," the bandit replied coolly, "This *is* my job."

Ken then shook the manager's hand and ducked down an alley to his rental car parked a few blocks away. He shipped the briefcase of bills by courier and arrived at the airport in time for the 5:00 PM flight back to Winnipeg. With the money, the twenty-six-year-old bandit bought presents for his children and a second-hand fox stole for his wife. He was, in fact, so filled with Christmas joy that he sent a card to Mr. Lunn, signing it, "Merry Christmas from a satisfied customer." Ken and Elva then flew the Stinson to

Mexico for a holiday, while Elva's mother babysat the kids.

The work and cost of developing the lodge at Moar Lake was never-ending. Every day, regardless of the weather, Ken flew supplies into territory that allowed for no flight miscalculations. He needed a financial backer, but the response was always "no." He had to get more money quickly to save his dream lodge, his home, and his hard-won community reputation.

Another final bank job would solve all his problems. He'd heard the odds: the first time a person commits a crime his chances for success are 80-20; the next time, the ratio drops to 50-50. Nonetheless, he purchased clothing to suit his new identity as Mr. McGill, a contractor from Welland. He packed the .22 automatic pistol into his briefcase and purchased a same-day, round-trip flight ticket to Toronto, traveling on St. Patrick's Day, March 17, 1958.

Mr. McGill from Welland secured an appointment at 2:00 PM with manager Howard F. Mason at the Bank of Commerce on the corner of Yonge and Bloor Street. As Ken accompanied the manager into his office, Mason gestured toward an open doorway leading to the accountant's office. "This branch has been plagued with robberies," he said, "so we've put a few safety precautions into place."

Ken nodded and weighed his chances. Then he said quietly, "This is a holdup."

"Police! Sound the alarm!" Mason shouted, then bolted as the accountant tripped the alarm. Ken ran outside with two bank accountants in hot pursuit. Someone stuck out a foot and tripped him. Ken fell, and the gun rocketed from his coat pocket. His capture was complete.

The news was blasted over radio and television, and in the newspapers. Although his family was devastated to discover what Ken had done, when Elva was interviewed by a reporter for *Weekend Magazine* she responded with calmness and dignity. "Why do you stay with this notorious man, this Flying Bandit?" the reporter asked, coining the lifelong nickname. "When you have loved and still love someone, you just don't drop them," Elva steadfastly replied.

Ken Leishman was brought to the Toronto city jail and treated with the utmost courtesy. In casual conversation, the guard mentioned the slickness with which Ken had pulled the bank job the previous December. "I was lucky that time," Ken humbly replied. He was caught in his own web. No longer a hero, Ken was thrown into Toronto's Don Jail.

On April 8, 1958, he pleaded guilty to one charge of armed robbery for the first bank job (receiving a sentence of nine years) and one charge of attempted armed robbery for the second (netting three years), with the sentences to be served consecutively. He spent one month in the Kingston Penitentiary before being transferred to the Stony Mountain Penitentiary near his home in Winnipeg.

Jail was not where Ken Leishman wanted to be, particularly since a sixth child, Robert Donald, had been born to Elva and Ken on October 22, 1958. However, other than escaping, parole was the only way out, so he became a model prisoner. He enrolled in high school correspondence courses from mathematics to French, and also qualified for a journeyman electrician's certificate. He joined debating and drama clubs; took Dale Carnegie courses; refereed sports; edited *Mountain Echoes*, the prison newspaper; wrote letters for other inmates; and accepted any job that was assigned. He exercised daily, honing his body down to a trim 187 pounds.

He wrote hundreds of letters to judges, politicians, social workers, the John Howard Society, and family members to plead his case for early parole. He especially consulted with Harry Backlin, a young law student he'd met when the university debating club had visited the prison. His energetic campaign worked. Ken received parole on December 21, 1961, after serving three years and eight months of his twelve-year sentence. He would remain on parole for 2,238 more days until his full sentence expired on March 28, 1968.

After his release, Ken met the branch president of the local Mormon church and the Leishmans began to attend. The family was baptized into the church on March 31, 1962. Things were on track once again. They had a new home and a Mercury station wagon into which the entire family could fit. A seventh child, Trent, was born on April 13, 1963. But as always, when Ken thought about it, his future seemed bleak. He was thirty-four years of age, working sixteen to eighteen hours a day as a cookware salesman, and going nowhere. He began to work out a daring new plan.

GOING FOR GOLD

The Winnipeg International Airport regularly received gold shipments from the mines at Red Lake, Ontario (Campbell Red Lake Ltd. and Madsen Red Lake Mines Ltd.). These shipments arrived on Sundays, Tuesdays, and Thursdays at 10:00 PM via Transair's Douglas DC-3 on Flight 108. The cargo was signed off, transferred to an Air Canada freight truck, and loaded onto a connecting flight for its destination, the Royal Mint in Ottawa. From a distance, Ken had observed the shipment as it came down the conveyor from the cargo hold of the aircraft—twelve wooden boxes at fifty pounds per box would equal six hundred pounds of gold. Based on Troy weight, this equaled 7,200 ounces and, at $35 per ounce, it was worth $252,000.

As his confidante, Ken chose Harry Backlin, now a criminal lawyer. Harry claimed to know overseas buyers who regularly dealt in such matters, and for 50 percent of the profit he agreed to finance the operation and arrange the sale. "Clean" helpers would be required to undertake the transfers, so Ken approached a perpetually broke salesman named John Berry and offered him "$10,000 for 10 minutes' work." John agreed and recommended a partner, another high-flying salesman named Richard (Rick) Grenkow, who was offered a similar deal. Then Rick told his brother Paul about the caper. Paul sold vacuum cleaners in Ontario and could hang around Red Lake as an on-site spy.

It was Harry who fell down on his commitment. He could only scrape up advance operating funds of five hundred dollars, which he "borrowed"

Ken Leishman with his children, on a trip to the Rockies. (clockwise from top left), Blair, Wade, Dale, Ron, Trent, LeeAnne, and Rob. (Courtesy Leishman Family)

from trust funds from a credit union where he was an officer. He sent Ken to collect it, thus linking their names. Harry's second surprise was to announce that he and his wife were planning a holiday, which would prevent him from taking an active role in the robbery. When Ken cut Harry's percentage to 40 because of his lack of financial, legal, and physical assistance, Harry readily agreed. Ken then asked John, Rick, and Paul to take an advance of one thousand dollars each, with the rest to be paid when the gold was sold. They reluctantly accepted.

For practice one night, they took the Air Canada freight vehicle, a white GMC panel van, for a spin when it was found in the hangar with keys in the ignition. No one questioned them. White coveralls and navy blue parkas were purchased from the Army & Navy Surplus store and hand-stenciled with Air Canada logos to resemble the company's ground crew uniforms. For hardware, Ken produced a .32 caliber Beretta.

Harry and Ken would require passports for the eventual overseas sale and gave each other's names as character references on their applications, although Ken was still on parole. Ken's passport came through on February 28, 1966, but Harry's was delayed because, he reasoned, he'd recently changed his name from Backewich to Backlin.

When Paul Grenkow spotted the boxes on the loading dock, he called to report that twelve wooden boxes were scheduled for the flight from Red Lake to Winnipeg. On Tuesday, March 1, 1966, Ken drove to the Winnipeg airport, met John and Rick who'd arrived in their Ford Galaxy ragtop, issued last-minute instructions, and drove off to an empty warehouse office near the airport to wait while the boys did their job.

John and Rick drove the Air Canada truck to meet Flight 108. They watched the cargo hatch open, the conveyor belt start up, and the boxes come sliding down. Then Rick handed the attendant a waybill and said, "This shipment has to be transferred right away—it's gotta get on a late flight leaving in thirty minutes." He and John helped to hoist the heavy boxes off the conveyor onto the dolly and into the opened back doors of the truck. Forms were filled out over the forged signature of an Air Canada official, the attendant added his, and Rick scribbled "Fred Davis" as the receiver. They secured the doors of the truck, and by 10:20 PM they were off with their load of gold.

Instead of following the usual Air Canada routine, however, they drove to the parking lot. While maneuvering the Ford Galaxy convertible closer to the open back of the truck, John hit a fence post with the open door of the car, knocking off the armrest. Once their mission was accomplished, they drove the tail-dragging convertible to meet Ken, and with the heavier than expected eight hundred pounds of gold weighing down the seats of his station wagon, Ken drove to Harry Backlin's house at 119 Balfour Avenue in Riverview. Harry's mother-in-law, who was staying in the house while Harry and his wife were on vacation, allowed Ken to store some "chemicals from the plant" in the freezer. After transferring the heavy load of twelve boxes, each weighing between forty-two and ninety-two pounds, he warned her against opening them because "air would destroy the chemicals."

By midnight, the St. James police airport security detachment was painfully aware that $382,436 worth of gold bullion had disappeared. They had no clues. Neither did the Winnipeg City Police, nor the RCMP.

When Ken phoned Harry at his vacation hotel in Los Angeles, Harry gave him the name and phone number of the "investor" in Winnipeg who would give him the $2,000 to pay the promised advance to the helpers. This was done.

When Harry returned from his holiday the following Saturday, March 5, he and Ken removed the gold from the freezer, unwrapped the bricks from the boxes, and stuffed them beneath the liquor bar in the basement. Panting with exertion, they put the last box, number twelve, into an old leather briefcase and hid it above a ceiling tile in the rumpus room. Then Harry related bad news: his local gold-buying connections had backed off. One of them would have to make a trip to Asia to sell the gold. Harry's passport hadn't yet been issued, and his court calendar was full. Ken was still on parole and not supposed to travel beyond the Prairies. Despite this, Ken was elected to make the trip from Winnipeg to Vancouver by train, on to San Francisco by bus, and to Hong Kong by air.

Ken needed a smallpox vaccination, so on March 7 he saw Dr. W. G. French and coaxed him into backdating the vaccination certificate to March 1 to accommodate travel rules. The last item on Ken's agenda was to report to his parole officer, which he did, commenting casually on the gold heist that was the talk of the country. With much trepidation, Ken boarded the train for Vancouver. Secured in a money belt about his waist was six pounds of gold that he'd hacksawed into three pieces from one brick to show potential buyers.

After checking for known criminals in the area, the police found Ken's record and quickly connected the Leishman and Backlin names. They visited Harry's office and laughed out loud when they spied an open briefcase containing thirty-six pounds of gold remaining from a sawed-off brick. Harry sheepishly led the officers to his backyard and watched them dig about in the snowdrifts until they found the rest of the treasure, which he'd moved outside from beneath his basement bar.

Meanwhile, Ken checked in at the Canadian Pacific Telegraph Office in Vancouver to receive the five hundred dollars (one hundred dollars less than promised) that Harry had wired for his flight to Hong Kong. He also discovered that Harry had booked his flight direct from Vancouver—eliminating the plan to hide his tracks by taking the bus to San Francisco—and that Harry had used his credit card to pay for the flight. Harry's stupidity had left tracks a blind person could follow. Ken decided to go home.

At the airport, Ken canceled his reservation and bought a ticket back to Winnipeg. When he arrived at the check-in counter he was told that the RCMP wished to see him. Before reporting to their office across the floor, Ken scuttled over to a pilot's supply store and bought a small sturdy box. He folded the money belt, stuffed it into the box, secured it with tape, addressed it to Harry Backlin in Winnipeg, and walked it over to the CP Air Cargo office. At least he wouldn't be caught with gold on him.

When the RCMP chastised him for breaking parole travel restrictions, Ken admitted his error and, on producing his ticket to Winnipeg, he was let go. Ken returned to the CP Air Cargo office, retrieved his parcel, and walked out of the terminal building. He came back inside with an empty belt.

The police were not through with Ken Leishman, however, and decided to retain him in Vancouver for further questioning. He was placed in a cell with an undercover cop, who easily got Ken talking about the recent gold heist staged in Winnipeg by someone obviously familiar with airports and aircraft. Soon Ken was revealing more than he should have about the job. He had been conned again.

The RCMP shipped him back to Winnipeg, where he was told of Harry's arrest. Several days later, John Berry and Rick Grenkow were also arrested— the accident with the Ford Galaxy hitting the airport fence post had given away their part in the maneuver—and Paul Grenkow was brought in for questioning. Initially, all denied any part in the heist.

At the preliminary hearing in Winnipeg in May, the undercover officer was able to repeat, almost verbatim, Ken's statements. The same man had also been placed in cells with Ken's three accomplices, and had performed his job very effectively: all had blabbed. Ken, John, and Rick were committed to stand trial on charges of conspiracy and theft. Paul was charged with theft, and Harry with unlawful possession of the gold bricks. Ken was denied bail and taken to Headingly Provincial Jail to await trial.

FLYING THE COOP

Ken Leishman's daring exploits, his stylish manners, and his eventual capture became the sensation of the nation. But he was miserable in Headingly, awaiting his trial scheduled for September 12. He was concerned for his family and about the length of time that might pass before he could be with them again. He faced the possibility of a ten-year sentence, plus the six years still remaining on his last sentence.

Six months of incarceration gave Ken plenty of time to think. He would escape and go to Cuba, where Castro was sure to be interested in his flying skills. Elva and the kids could meet him in the United States and they could all fly away, forever. Ken memorized the jail's passages, doors, key and lock systems, and the number of guards and their backups. Then he carefully chose a partner: Joe Dale, a handsome, quiet, First Nations man who was doing time for armed robbery and rape.

The escape was timed for 7:00 PM on September 1, 1966, because the night shift retained only a skeleton crew. "We were not going to simply escape and take off leaving the jail staff armed and on our heels," Ken wrote in his journal. "This would be a prison take-over."

First, they alerted the guard by yelling out that one of the prisoners,

Barry Duke, considered criminally insane and committed on a murder charge, was having a fit. When the guard came to investigate, the inmates pounced, relieving him of his keys and weapon. Ken and Joe took the lower tier, while other inmates took the upper one, overpowering guards, locking them in empty cells, and taking their weapons and keys. George Leclerc, a Montreal man in for theft and false pretenses, and Barry now carried .22 caliber revolvers, while Joe packed a heavy .45. In the basement, the four men found their own clothing, hung in order and labeled. They quickly pulled on pants, shirts, and shoes, released a third set of bars, and were gone.

Once outside, Ken Leishman, Joe Dale, Barry Duke, and George Leclerc jumped into a Chevrolet Bel Air belonging to the prison and spun down the gravel driveway. The time was 7:15 PM. Eleven guards had been locked up, while fifty prisoners milled about not knowing what to do, some afraid to make the final break. Five other prisoners, besides those in the Chev, scattered on foot into the night. No one had been injured or killed, no shots were fired, and those who wanted to escape were given the opportunity.

After ditching the marked car, Ken and his three pals ran through fields until they came upon a farmhouse near Stonewall, twenty miles from Headingly. Upon receiving no answer to their knock on the door, George Leclerc broke the lock with a sharp kick. They entered and secured the broken door with a bandanna. Fortunately, the house was unoccupied, with the furniture draped in sheets. On the kitchen counter were some fresh tomatoes and a loaf of bread, indicating that a daytime visitor might come by to water the livestock and eat a quick lunch. There were no dishes or cutlery, but there was some food in the freezer. While the other men found beds and dropped into sleep, Ken, equally exhausted, kept watch in a chair.

The next morning, an older man driving a tractor and hauling a cultivator pulled into the farmyard. He watered the cattle, then went round and round cultivating the garden as the armed men watched nervously from behind the curtains. Finally he left.

Later that day, Ken was asleep when he heard the doorknob rattle, then stop. Grabbing a gun from the counter, he went to the window and peered out from behind the curtain. A car was parked in the driveway with a woman sitting in it, while a man stood at the door. Ken woke the men, who remained cautious and quiet until the car finally roared down the driveway and onto the road. They knew they'd been discovered and immediately set off walking across the darkening fields.

Soon they encountered a young couple in a parked car, who weren't paying attention to anything outside the vehicle. The bandits jumped into the car while Ken soothed the frightened couple: "You have nothing to worry about if you just keep quiet and come along for the ride." They did.

The best escape would be by air. Being familiar with nearly every small agricultural center from his days of flying around repairing and selling farm machine parts, Ken remembered that there was an airport just outside Steinbach. He knew that the town's predominately Mennonite population of seven thousand would likely be at church this Sunday morning during the Labour Day long weekend. Steinbach was also ideally located about thirty miles south of Winnipeg, near the Manitoba/USA border.

At the airport, Ken found a four-place Mooney Mark 21, new and ready to go. He started the aircraft engine by breaking the instrument panel (inadvertently also damaging the radio) and hot-wiring it, then they lifted up the hangar doors and pushed the airplane outside. By 7:50 AM the Mooney was airborne. Ken circled, waggled the wings at his young friends on the ground, and, like Canada geese, the bandits flew south.

Following a road map, the escapees flew at low altitude across the border—destination Cuba. When the Mooney got low on gas, a small airstrip was selected at Tyler, Minnesota. The pumps weren't operating, but the attendant offered to buy them breakfast at the local restaurant. They accepted, as they were famished, reeling from lack of sleep, and Barry's mental condition was rapidly worsening. Their elation over a free meal turned to dread when the local newspaper reporter arrived to take photos and get a story. There was no way out, except to pose around the airplane as if working on it, in positions that they hoped would hide their identity.

Back home, the police had received a call from the young couple whose car was hijacked and who—oddly—had waited the thirty minutes requested by Ken. The police were confident that stories would soon surface about a tall, handsome, friendly, talkative pilot accompanied by three ne'er-do-wells, flying around in a maroon and cream Mooney. The Flying Bandit wasn't exactly invisible, or invincible.

After leaving Tyler, Ken made a stop for gas in Springfield, Minnesota, seventy miles to the southeast, "landing on fumes." He purchased aviation maps of the lower eastern states and headed toward Chicago, flying along the western outskirts of the city. They'd need new identities, and Ken suddenly remembered that a friend in Winnipeg had a brother named Paolo who lived in Gary, Indiana. He could find safe houses and direct them to helpful people who sold passports and anything else they'd need.

Their next landing was in a farm field near Gary. The men proved so amiable that the farmer, Russell Shook, and his wife invited them in for coffee and food, then drove them to nearby Hobart, where they bought bus tickets for the short trip to Gary. They entered the downbeat M&S Tavern in Gary and inquired of the bar manager, Christ (Chris) Stath, if he knew Paolo. "Yes, I've heard of him," he replied cautiously. On the television news he'd also heard of the escape of four prisoners from Canada. Maybe there was a price on their heads. He slyly recommended the Balmoral Hotel across the alley where they could get cheap rooms and wait for Paolo to be contacted. The Shook family had also seen the news, which carried a description of the airplane still parked in their field. Russell Shook called the local constabulary. The RCMP and FBI were subsequently alerted and went into action.

The takedown came as a complete surprise to the escapees. Police cruisers surrounded the hotel, spotlights glared into the windows, doorways, and even over the rooftop. Suddenly the door to Ken and George's hotel room was thrown open and a man in civilian clothes stood in the lighted hallway, pointing a .38 revolver into the room. "Come out of there," he ordered. Ken fingered the gun he'd set on the bureau and thought how easy it would be to shoot the gun out of the man's hand. George sat on the bed back in the shadows, eyeing his gun, which sat about three feet away. "Turn on the light in there!" the armed man ordered. As George reached with one hand for the string to pull on the light, his other hand went out along the bed reaching for the gun. He fired, and instantaneously leapt in the direction of the window. There was an immediate volley of gunfire from the hallway.

Ken yelled, "If you want to shoot, go ahead, but let me out of the line of fire!" He then left the room with his hands raised to his ears, as armed, uniformed police swarmed the room. The police had ambushed Barry in his room down the hall, while Joe had made a run for it. Now, Barry and Ken were handcuffed and taken down to one of the dozens of police cars parked helter-skelter in the street.

A huge crowd had gathered to watch the show, even though shots rang out as George and Joe tried to escape over the roof. George attempted to cross over the lane via a twelve-inch pipe some thirty feet above the ground, hanging from the pipe while Joe waited for him to reach the far side. Suddenly the police were everywhere, and Joe had no choice but to run across the pipe like a tightrope artist. He then turned to fire two shots over the heads of the pursuing police squad. He and George smashed a window in a building, severely cutting George's arm. Joe stopped just long enough to take off his belt and use it as a tourniquet on George's arm, which was pouring blood, before leaving George inside the building and escaping once again out onto the roof as tear gas was fired into the building. Gunshots exploded from sixty or so armed policemen in a scene worthy of a Clint Eastwood flick—twenty rounds of ammunition by later count—while George hid behind a tank on the roof. There was no escape. "It was a case of surrender or die," Ken recalled. "The police were too eager to shoot at anything that moved."

En route to the police station, Ken kept up a patter with the police, as well as with onlookers who crowded around the cruiser. Even though their daring cross-border flight had netted them just fifty-two hours of freedom, Ken's legendary role as the Flying Bandit was assured. The escaped prisoners were flown back from Windsor to Winnipeg on September 8, 1966, on an Air Canada Viscount, and greeted at their home airport by a throng of thousands. There was just time for a smile and nod to the crowd, and a much-quoted statement from Ken: "We're no mad-dog killers!" Elva, with three-year-old Trent and Ken's mother, sat watching from the *Winnipeg Free Press* car, as close as they were allowed to come. The four men each faced six charges, and Ken Leishman still had to answer to charges pending from the gold heist.[1] He was going to have a long vacation as a guest of the Feds.

On October 27, 1966, Harry Backlin appeared in court to answer charges of possession of stolen property, and received a sentence of seven years for his part in the scheme. Ken's Fly-In Fishing Lodge on Moar Lake was repossessed that year by the government for back-taxes and was subsequently sold to a conglomerate of Winnipeg newspapermen.

THE CELL FROM HELL

Ken was the only prisoner in the basement eight-cell solitary confinement unit at the provincial detention center on Vaughan Street in Winnipeg. He was told that his new home had been named the "death cell" by cons who'd survived it, in honor of those who hadn't. A twenty-four-hour guard was posted outside his cell—as if anyone could escape through the small steel-barred door or tiny windows encased in the thick stone walls. Ken chose this time of physical inactivity to write his memoirs, but revealed nothing that might jeopardize his precarious legal position at the upcoming trial.

Ken wasn't allowed even a daily exercise walk until more than a month had passed—his only walk was to appear in court—and his deteriorating health became a public shame. He was subsequently given permission to exercise twice daily in the hallway outside his cell. During these short walks,

Ken discovered a stone exterior wall at one end of the hallway and a steel mesh door at the other end. The door had two-inch-square criss-cross openings and was secured by a heavy pin-and-bolt locking system, which Ken studied with calculated interest. It was just a mechanical device, and Ken had always been handy.

He found a two-foot length of electrical wire from an unused bell system in a crawl space between the ceiling and the top of the cell, and deftly fashioned it into a noose that ended in a spiral cone. From this he suspended a cloth pouch made from a corner of his bedsheet that he had bitten off, and hid the apparatus in a mouse hole (remaking the pouch when the mouse ate it). Noise from the radio and Ken's frequent coughs muffled the sounds of his blanket being ripped into strips, with which he planned to tie up the guard. A twelve-inch length of cast-iron plumbing pipe was also found and hoarded.

On Sunday, October 30, during his exercise period, Ken quietly slid the pouch and wire apparatus into the door lock and hooked the wire over the bolt end, causing the pouch to drop over the pin. When he pulled up on the pin, the bolt slid back and the lock released. He left it, went back to his cell, and gathered Elva's letters and his manuscript into an envelope he then hid under his clothes.

Coughing furiously, Ken asked the guard, Paul, to call down Lloyd, one of the senior officers. Ken had spoken with Lloyd before so it wasn't an unusual request, and Paul complied. Then Ken's plan sprung into action: he ordered Paul to the end cell under threat of being clouted with the pipe, and when Lloyd came down, Ken took his keys and threw him into the end cell as well, tying him up with the blanket strips as he lay on his stomach on the floor.

Ken then ordered Paul to summon a third guard, George, who ran downstairs to meet the same fate. A scramble ensued, but Ken escaped, ran up the stairs, grabbed a heavy brown parka, and shoved the key into the last lock, which brought him outside to the exercise yard. The adrenaline rush helped him to clamber over a twelve-foot wooden fence topped with strands of barbed wire. As he leapt to freedom, a piece of barbed wire ripped open his hand, causing excruciating pain and a telltale trail of blood. His ankle, sprained from a hard landing, throbbed unmercifully.

Ken half ran and half walked across a two-hundred-yard open area near the legislative buildings directly to the south and west of Vaughan Street, terrified of being spotted by the police cruisers that must now have been converging on the area. He headed toward his mother's apartment building, but although he could see her light on, he couldn't take the chance of going to a place where the police would be sure to check. Six blocks further west, he reached Portage Avenue, the main east-west thoroughfare through the heart of the city. He crossed Portage and leapt to the security of a laneway, then angled west across Sergeant Avenue, through the warehouses along Henry Avenue, to a few blocks south of the CPR yards. Small houses in small fenced yards severely impeded his progress, as did the high chain-link fence surrounding the CPR yards. Ken's foot was throbbing and dogs were barking. He encountered some kids and an old woman who looked suspiciously at him, this bearded man in the hooded parka. He clambered up the high locked gate enclosing the yards, swung over in a single motion, and hit the ground rolling. Now he had to cross between lines of railway cars filling track after parallel track.

Two hours later, near 9:00 PM, he arrived at his destination: a house occupied by reliable friends. They were not home. He decided not to break in—maybe they'd moved and new people occupied the house—and limped on as police sirens wailed. He couldn't recross the CPR tracks; he'd have to continue north or east. Elva had relatives in West Kildonan, two or three miles distant, who surely wouldn't turn him in. He decided to head toward Main Street, travel parallel to the west of this busy thoroughfare, and find a phone booth in a darkened corner. He walked and walked until nearly 11:00 PM.

He made it to Main Street and Jefferson Avenue, where, in the corner of a supermarket parking lot, he spied a phone booth. He waited in the shadows until two men in the booth finished their call and left in their car. Now what? He wanted to call Elva, but the police might be there, waiting for his call or arrival. He phoned his lawyer, asking him to come.

As he exited the phone booth, he came face to face with Constable Ed Finney of the West Kildonan detachment, who held out a pair of handcuffs. The two men who'd used the phone booth directly before Ken had heard on their car radio of an escaped convict wearing a brown parka, and recognized him. When they'd banged on the door of a neighboring house, crying, "We found Ken Leishman! Call the police!" the lady of the house had replied, "Why don't you leave the poor man alone?" Many Winnipeggers agreed, to the chagrin of the police and law-abiding citizens who did not feel similarly charmed by their hometown outlaw.

Court day, November 1, 1966, found Ken Leishman once again looking jaunty in a smart suit, dress shirt, and tie. He pleaded guilty before Magistrate Gyles to all charges and was sentenced to five years for the escape from Headingly Jail, plus two years each for the Vaughan Street detention center escape and other charges. The judge ordered that the two-year sentences be served concurrently rather than consecutively, however, for a total of seven years. Later that same day, Ken faced Judge C. I. Keith on theft and conspiracy charges stemming from the gold heist. Ken was given a total of eight years—resulting in only a one-year sentence for the gold robbery. It was a break, but not much.

While the people cheered, the prosecution did not. Neither did Ken, or Elva, who was ever-present in the courtroom. Eight years is a long time for a family to manage on its own, and Ken had seven children whom he wouldn't see for their most important years. A further blow came on January 17, 1967, when Ken was ordered by Magistrate Wallace Darichuk to also serve the six years remaining on his twelve-year sentence for the bank robberies.

In May 1967, John Berry and the Grenkow brothers pleaded guilty to possession of stolen property and each received three-year sentences.

Time and Time Again

The story of the Flying Bandit sparked international interest. Hollywood actor/producer Darren McGavin, then starring in two popular television series, *Mike Hammer* and *Riverboat*, came to Winnipeg and paid Ken five thousand dollars cash for a film option on his autobiography with the promise of production and royalties to follow. However, nothing further came of it and the manuscript was never offered elsewhere for publication or production.

While serving his time, Ken found emotional release in writing letters on behalf of himself and other inmates, and studying. He formed a drama club, read prolifically, and wrote poems to his wife:

> I never was a husband of any great acclaim,
> I often wonder, sweetheart, how I sold you on my name.
> But the years fulfill their purpose, and your husband's
>   come of age
> Some men mature in battle and some within a cage . . .

Ken also studied for his high school matriculation while working in the jail's library. In September 1967, he was transferred from Stony Mountain to the maximum security penitentiary in Prince Albert, Saskatchewan. His education was interrupted and, most importantly, visits from his family ceased because of the distance. Through arduous effort, he passed his matriculation exams in June 1968, but it was a low time. Lee Anne was married in 1969, but Ken was unable to escort his only daughter down the aisle.

When Ken heard that his old cohorts Harry Backlin and Rick Grenkow had been granted parole to attend university, he decided to pursue higher learning as well, as he would be eligible for parole in June 1969, after serving one-third of his eight-year sentence. He enrolled in correspondence courses in computers, and an economics course from Queen's University, which he passed the following April. He then applied at Red River Community College in Winnipeg to take courses in engineering technology beginning September 1969, and he continued to write letters begging for a transfer back to Winnipeg and a parole hearing.

When Elva filed for divorce in April 1969, Ken's urgent pleas for a transfer finally gained sympathy. With half of his eight-year sentence completed, he was transferred back to Stony Mountain in June, and Elva halted the divorce proceedings. He was granted weekend passes to visit his home, and day parole, which allowed him to personally register for college on September 8, 1969. The Leishmans' eldest son, Ron, was also attending Red River Community College in the advertising art department, while Ken entered through the engineering door. Ken soon joined the college's flying club and was elected president. He applied for and passed his medical, and his pilot's licence was renewed. Job offers started coming in, pending his parole hearing.

On January 9, 1971, Ken received permission to take his family on the one-hundred-mile drive to visit his ninety-year-old grandfather in Treherne. When a storm hit, making road travel difficult, Ken, Elva, and Trent instead drove forty miles to Steinbach to visit a potential business partner. Following discussions, Ken was offered the keys to the company's Piper Super Cub to take Elva and Trent for a ride.

On January 13, six days before he was due to write his final third-term exams, Ken's day privileges were canceled and his parole hearing deferred for two years due to his unauthorized flight. Ken wrote his exams under penal supervision, and passed. Then came the long struggle for another parole hearing.

At this point Ken decided to give his records a forensic audit. He calculated the time sentenced versus the pluses and minuses of time spent in incarceration, parole permissions and revocations, good behavior subsidies, and seventy-two days spent "detained illegally" while awaiting various trials, and his math proved that he had more than served his sentence. He presented these facts in an appeal to the Federal Department of Justice, which recognized their correctness and released him on May 3, 1974.

FREE TO FLY

It was time for Ken Leishman to decide what to do with the rest of his life. Elva had stuck by him through the worst years; now he vowed to give her some of the best. Airplanes had been an integral part of Ken's life, and bush flying particularly appealed to his love of aviation and desire for freedom. Using money both borrowed and inherited, Ken gained his commercial pilot's licence. His first job took him, Elva, and Trent to Sioux Lookout, Ontario. There, Ken volunteered for everything from hauling cargo to medical evacuations, flying in to small, almost nonexistent, airstrips to gain experience and flying hours.

Ken jumped at an opportunity to manage, and eventually buy into, Tomahawk Airlines in Red Lake, Ontario. The fact that Red Lake was the source of the gold he'd heisted in 1966 from the Transair flight was just an odd coincidence. The gold-filled flight was still being made twice a month, but that didn't interest him anymore. Owning an airline was a chance for a real, secure future. Unfortunately, the opportunity collapsed, but Red Lake still offered lots of work and the residents welcomed the Flying Bandit and his family. Ken started flying on contract where and when he could.

One day in 1977, his aircraft had just been loaded with passengers, and Elva, as usual, went around to the back door to ensure that it was secured for takeoff. "I looked in at Ken in the cockpit and knew something was terribly wrong—his eyes were going funny," Elva recalls. Ken spent eight days in the hospital, diagnosed as suffering some kind of seizure. He lost his pilot's licence for two years. He and Elva then opened a sporting goods and clothing store and called it "The Trading Post." The store provided a living and quickly became a center where tourists flocked to meet the still-famous

Ken Leishman with Taylorcraft. (Note the Champion wind generator under the belly.) (Courtesy Leishman Family)

Flying Bandit. In January 1978, Ken Leishman was elected president of the Red Lake Chamber of Commerce. Encouraged and interested, he ran for the position of reeve of the town council, losing by only a few votes. He built a houseboat, and when his health recovered, took relief flying jobs with Sabourin Lake Airways. He was living the good life, finally.

Then came the fateful flight of December 14, 1979, when Ken went missing.

THE DISCOVERY

The search for the Aztec piloted by Ken Leishman, reported missing on December 14, 1979, resumed on April 30, 1980. The weekend of May 3 and 4 marked a "point of decision" for the search and rescue effort, whether to continue this seemingly hopeless and increasingly expensive search or call it quits. On Saturday, May 3, 1980, shortly before 10:00 AM, the lost aircraft was spotted twenty-five miles northwest of Thunder Bay, precisely six miles north of Finmark, in an area covered with thick woods and dotted with small sloughs. Canadian Armed Forces and Lakehead Search and Rescue Unit search planes had passed over the area nineteen times, and Ontario Provincial Police flights six times, without spotting the wreckage. The discovery marked exactly six years to the day from Ken's release from Stony Mountain in 1974.

A swath slashed through the trees told the story of a rapid and fatal descent. One large tree was clipped at the sixty-five-foot level, with four smaller trees and brush sheared for 460 feet before the aircraft had hit a large rock outcrop, bounced up, and continued for another one hundred feet before smashing into the trunk of a large evergreen tree. The front section of the Aztec's fuselage had been ripped off, a wing was wrapped around the passenger area, and debris was scattered over a three hundred by one hundred foot rectangular area. The impact had snapped the seat belts and sent the seats and Harper's stretcher flying out the front. In the opinion of the experts, no one could possibly have survived the crash.

The bits and pieces of fleshless bones found scattered over a half-mile radius ("about enough bones to fill a man's cap," Elva says) were sent to the Centre for Forensic Sciences in Toronto, which concluded only that they were from humans. Positive identification seemed impossible without references such as teeth to match dental records, or a portion of a once-broken and healed bone to compare with medical files.

Evidence found near the site showed that the bodies had been devoured by animals during the five months the wreckage had remained undiscovered, and cloth found in nearby animal excrement matched the type of clothing worn by the victims. Ken's wallet was found and returned to Elva, bearing animal tooth marks, but still containing his credit cards and personal receipts. One of Ken's torn boots was discovered in two pieces; its crepe sole was near the aircraft while a portion of the leather upper had been taken a quarter-mile away by an animal. In an effort to find further evidence, a beaver pond located two hundred feet from the wreckage was dynamited and drained, and searchers sifted through the muck. Some money and Leishman's personal papers were discovered at the bottom of the slough, along with a long-dead moose carcass.

A memorial service was held in Winnipeg on May 9, 1980, in the Mormon faith. The newspapers also gave Ken an honorary send-off, listing his positive achievements while omitting his criminal past. Rather, they chose to quote from one of Ken's poems that revealed his feelings for flying the bush:

> . . . this is a land of special beauty, it's a land for special men.
> When I leave I'll do so gladly but I know I'll come again.
> I'll bear memories of kind people, of sunsets without end,
> I'll respect and fear the Northland, and I'll do so as a friend.

In July 1980, burial warrants were issued and the few remains found of passengers Eva Harper and Janet Meekis were released so the First Nations band at Sandy Lake could conduct ceremonial burials. In October, Thunder Bay coroner Dr. Douglas Rathbone stated that, although the lab was still unable to identify Ken as the third victim, an inquest should be held where a jury could decide whether or not he should be declared dead. Rathbone then turned the case over to the northwestern Ontario regional coroner, Dr. William Wigle of Dryden.

A dilemma now confronted Dr. Wigle. How could a person be legally declared dead when there was no identified body and no real evidence that death had in fact occurred? Lawyers were waiting to settle the estate and banks to close their files, but in Dr. Wigle's view it might be necessary to search again in the fall for any clues that might prove Leishman had truly departed this earth. Could Ken Leishman have lived through this horrendous crash? Rumor persisted that he had, but this was denied by those who viewed the site at close quarters. Corporal Jim Solomon of the Ontario

Provincial Police told the *Thunder Bay Times-News* it was obvious that animals had eaten the victims: "I've seen large deer that have been attacked by wolves and all that was left was a tuft of hair." Further, there was no sign of any human activity around the crash site that indicated survival, such as a campfire, shelter, or signs of movement.

Nine months after the crash, the coroner's office still officially listed the pilot as being alive because no human remains could be identified as Leishman's. What was to be done about this enigmatic man? The public didn't want to let go of the idea that Leishman, who in many people's minds was a thoroughly likable individual, might once again have evaded "capture." Newspaper reporters flaunted his titles: "flying bank bandit, gold robber and escape artist —a bona fide Northern hero." "Even the police liked him," one newspaper report stated. No one wanted him to be dead, neither his large family, his extended circle of friends, nor the press—and so he was kept alive through media coverage and hope.

At the coroner's inquest held on December 16, 1980, Dr. Thomas Heringer acknowledged Leishman's "rather exciting background." "He was sort of a well-known citizen and many are reluctant to believe he died in this plane crash," Heringer said, but added, "It is the duty of this inquest and the jurors to decide if he is dead or can be presumed to be dead." The jury's conclusion was that Kenneth Leishman, Eva Harper, and Jacqueline Meekis had all died from multiple injuries when the plane, traveling at 225 miles per hour, came down in rough terrain in high country about twenty-five miles north of the city of Thunder Bay.

"Crash locators called for; Leishman ruled dead" reported the *Thunder Bay Times-News* on Wednesday, December 17, 1980.

> A rumour has persisted since the crash, due mainly to Leishman's colorful history, that he perhaps planned the whole thing and managed to get out alive to retrieve mythical millions in stashed gold and money. Wade Leishman (son) said there is no

basis for believing such stories. "He had nothing to gain. He had no life insurance and no motive. I believe my father died on December 14." He said later that no company was willing to insure his father because of his past penchant for unusual lifestyle.

The last entry in the Aztec's log book, in Leishman's handwriting, stated: "Dec. 14. Destination YQT Thunder Bay. Pilot Leishman. Three aboard. Up 15:05" There was no entry in the "down" column.

And so Ken Leishman's action-filled life story came to a full stop.

REFLECTIONS

Since then, the Leishman family has scattered over wide distances. The children have done well, certifying the steadfast care Elva gave them during very trying times, and the unsteady, somewhat controlling, but nonetheless loving attention of their infamous father. Elva counts her children as her "jewels," and she and her second husband (until his death in 2004) regularly visit them, her twenty grandchildren and eleven great-grandchildren ("and more to come," Elva adds). They, in turn, have survived the notoriety of their father to live successful, albeit more quiet, lives. "Mom and dad wrote to each other every single day while he was in prison," son Ron recalls. "But after dad was released, he lit a bonfire and threw the letters all in. Those letters would be great to have now. They chronicled our lives in minute detail."

Others whose lives were affected by Leishman's activities also remain philosophical about his exploits. Nes Ewanek, now of Calgary, whose Stonewall farmhouse hosted Ken and his fellow escapees, thankfully recalls that his family was in Alberta when the men invaded his house. But he shudders to think how close his father, Oliver, came to danger when he arrived on his tractor to water the cattle and cultivate the garden. Even more alarming was the arrival of his brother Eugene with his wife sitting in the car, who had noticed the bandanna that held the broken door closed. But Nes Ewanek harbors no grudge: "I guess you've got to do what you've got to do. They were desperate men. I think Leishman kept the others under control. He was the smartest, the most sensible of the lot." In fact, Ken Leishman was reported to have an IQ of 146—bordering on genius—no surprise to those who knew him.

Epitaphs abound on this enigmatic person. Heather Robertson states in her book, *The Flying Bandit*, that had Ken Leishman gone straight, with his intelligence he could have become a corporate president. Instead, he went crooked and became a legend. In the last act of a stage play written by Lindsay Price, also called *The Flying Bandit*, the contradictions of Leishman's life are summed up: the bank manager he robbed thought he was well educated, yet Ken had only a grade seven education at the time; he was thought to be a prosperous businessman, when in fact he was broke; he was referred to as a dapper gentleman, but he was a fugitive from justice.

The six pounds of gold that Ken "secured" in an irrigation ditch at the south side of Vancouver International Airport has never been found. "Ken told me he'd thrown the three pieces of gold into secret places outside the terminal building," Elva says. "He laughed later when he learned the area had been cemented over during construction."

Ken Leishman had a vision, but the shortcuts he took to achieve his goals too often sent him on the wrong road. But the legend of the Flying Bandit still thrives—on paper, on stage, in Ken's large and successful family, and in the minds of those who love a daring story.

STALIN'S FALCONS

Sigismund Levanevsky, often referred to as "Russia's Charles Lindbergh," was a bona fide hero of the Soviet Union and understandably chosen to pilot a pioneering flight over the North Pole on August 12, 1937. Levanevsky and his five-member crew took off from Shelkovo Military Aerodrome near Moscow at 18:15 hours in a four-engine DB-A (long distance bomber-academy) transport aircraft bearing USSR registration N-209. Fairbanks, Alaska, would be the initial stop in a flight taking them north from Moscow to the pole, then south along the 148 meridian over the pole and Canadian territory to New York. Their cargo included mail, furs, a barrel of caviar, and something that has been called a "secret load."[1]

Flight headquarters in Moscow anxiously awaited radio contact. The first few messages revealed no problems. Then, after the aircraft passed the North Pole at 14:32 hours (Moscow time) on August 13, flight headquarters received signs of distress: "Extreme right engine out of order. Oil pipe broken. Altitude 4,300 metres [14,000 feet]. Thick clouds. Wait . . ." The broadcast faded and N-209 was lost.

The disappearance of the big bomber and its crew somewhere over the Arctic shocked the world. The famous American pilot, Amelia Earhart

Putnam, had mysteriously vanished just six weeks earlier on July 2, while flying a Lockheed 10 over the Pacific. These losses dealt devastating blows to two major aviation pioneers, the United States and the Soviet Union. It was imperative that the aircraft and their crews be found, at all costs.

THE MAKING OF A SOVIET HERO

Sigismund Aleksandrovich Levanevsky was not only a hero to the Soviets, but had also established a worldwide reputation as a daring and competent pilot. He was blond, blue-eyed, and handsome, spoke English (which he had learned during visits to the United States), and was endowed with sufficient arrogance, ambition, and self-confidence to fit the image of an internationally acclaimed aviation hero.

Levanevsky was born in Sokulka, Poland, in 1902, and had always dreamed of becoming a pilot, especially in the North. "Airmen were the darlings of Stalin's regime, the embodiment of heroism and of the 'New Soviet Man,'" writes Von Hardesty. "They normally flew in teams, an expression of the collective spirit of the Revolution. *Stalin's Falcons*, as his fliers were called, in turn dutifully served the cult of personality." [2]

In the spring of 1922, Levanevsky attended a military school for marine fliers at Sevastopol on the Black Sea. In three years' time he became head of the flight school in Nikoleyev, and a year later principal of the All-Ukraine aviation school in Poltava.

A rescue mission in June 1933 made Sigismund Levanevsky famous. Jimmy Mattern, an American aviator, was flying along the coast of eastern Siberia from Khabarovsk en route to Anadyr in an attempt to make a solo circumnavigation of the world in his Lockheed Vega. When Mattern was reported missing from what the *Los Angeles Times* called a "world-girdling hop," Levanevsky was sent to search for him. Mattern, with his Vega aircraft down, damaged, and unflyable near Anadyr, was "extremely happy" to see his Russian rescuers. "We put Mattern ashore in Nome, Alaska," Levanevsky states. "He fell to the ground and began to bang his arms on the ground crying, 'America! America!' He was a happy man." Levanevsky concludes, "I know this very well. In one's motherland the snow is somehow special, and the air different, and the stars shine differently."

On February 13, 1934, Levanevsky heard a radio report that brought him to his feet. The icebreaker *Chelyuskin* had been crushed by ice. The ship and hundred-man crew—headed by Otto Yulevich Schmidt and called the "Chelyuskinites"—had become stranded on an ice floe near Chukotka. Moscow sent pilots Sigismund Levanevsky and Mavriki Slepnyov, along with polar explorer George Oushakov (the government commission representative assigned to the rescue of the Chelyuskinites) to America, believing it would be quicker to reach the stranded ship from the Alaskan side.

After spending ten days in New York before receiving orders to proceed to Cape Vankarem, the two aviators flew to Fairbanks, where they accepted two modified Consolidated Fleetsters Model 17AF (high performance passenger aircraft), NC-703Y and 704Y, from Pacific-Alaska Airways, complete with American air mechanics to assist them on the journey north.

Aided by Arctic-based radio operator Ernst Krenkel, who was on the ice with the Chelyuskinites, Levanevsky took off from Fairbanks on March 26 in poor weather to find Schmidt's camp. It was a flight from hell. After being forced down by weather for two days at Nulato on the banks of the Yukon River, they proceeded on to Nome. En route, they were caught in a snowstorm. A forced landing tore the skis off the aircraft and nearly killed its occupants. Levanevsky's mission was unsuccessful, and he and his crew were ignominiously transported to Vankarem by dog team.

Even though six other Russian pilots actually made the successful rescue of the Chelyuskinites, Levanevsky found himself included in Josef Stalin's presentation of the title, "Hero of the Soviet Union." He was also awarded the Order of Lenin. From that time on, these men were revered. His humble acknowledgment that he was not so deserving as the others gained praise from the highest order when Stalin shook his hand and asked, "Levanevsky, why are you always hiding yourself and are so modest?"

In 1935, Soviet artist V. Zavyalov produced portraits for postal stamps depicting the first Heroes of the Soviet Union, adorned with laurel branches forming wreaths of honor. Ironically, in Levanevsky's portrait one of the branches has been identified as myrtle, denoting death. That same year, Levanevsky invited pilot Georgi Baidukov to his home. As his wife Natasha, son Vladik, and daughter Eleonora tiptoed about the luxurious three-room apartment, careful not to disturb the conversation, Levanevsky described to Baidukov the wonders of the Arctic and, in particular, his own desire to do more northern flying. His excitement convinced Baidukov to join him in a plan to fly over the North Pole, to prove the feasibility of commercial air service between the Soviet Union and North America.

The historic flight was set for August 3, 1935, and brought worldwide excitement to aviators. On that day, Levanevsky, co-pilot Baidukov, second pilot Gurevich, navigator Levchenko, and reserve navigator Reliakov took off from Moscow for San Francisco in a single-engine ANT-25 aircraft designed by Andrei N. Tupolev. Over the Barents Sea, after ten hours in the air, oil leakage was reported. Baidukov wanted to continue, but Levanevsky said no, and they turned back.

It was discovered that the leakage was caused by overflow spillage through a drain pipe, a "constructive shortcoming" in the design, but not a serious matter. Baidukov had been right—to continue would have brought everlasting glory. But Levanevsky blamed the design of the aircraft. Tupolev was shamed, and Levanevsky's decision vindicated. Baidukov later reported: "In my opinion Levanevsky spared very little time in preparation for this flight. He familiarized himself very superficially with the aircraft and did not particularly want to get to know it. For him, the main thing was to achieve the set goal, but never mind in what way."[3]

In the spring of 1936, Levanevsky and co-pilot Viktor Levchenko visited the United States to check out that country's progress in engineering and aviation. "In America the idea came to me—to make a flight from Los Angeles to Moscow and to study the aerial route from the USA to the USSR," Levanevsky writes in his memoirs. "The Government of the Soviet Union permitted me to accomplish the flight. From Moscow they communicated that it was necessary to purchase an American aircraft. I chose for the flight one of very reliable design, the Vultee [V-1AS]. At the Vultee factory, near Los Angeles, the aircraft was built and fitted out within one and a half months. According to our directions they eliminated some construction short-comings."

Their plan was to fly on floats from Los Angeles along the west coast to Alaska, and then to Krasnoyarsk in central Russia. With worldwide good wishes, they took off on August 5, 1936. In Canada they hit a band of fog and rain, which persisted until they were forced to make a landing and spend a miserable night on an uninhabited island. At daybreak they brought the aircraft out at high tide and headed for Swenson Bay, forty miles east of Seward. There they received a telegram from Loy Henderson, the American charge d'affaires stationed in the Soviet Union:

> On behalf of the American people I am glad to pass hearty congratulations to Levanevsky in connection with the successful accomplishment of this flight . . . I am convinced that the route explored by Levanevsky will subsequently become not only a bridge for the joining of two continents, but a means of helping friendship between the great nations.

They'd made it from the USA to the USSR, but in an American-made aircraft and not over the pole.

In May and July 1937, two successful flights were made from Moscow to the United States—but not by Levanevsky. The second flight, piloted by Mikhail Gromov with co-pilot Andrey Yumashov and navigator Sergei Danilin, set a world nonstop long-distance record of 6,305.7 miles in sixty-two hours and seven minutes. At this point, Levanevsky's pride came into play. He *must* make a successful transpolar flight, but it would not be in an ANT-25 that had, in his view, let him down in his quest to be the first to fly over the North Pole to North America. He set out a plan in a letter to Josef Stalin and received the assignment to pilot an aircraft designed especially for the flight from Moscow over the pole to Chicago and New York.

The American-based Soviet official who had arranged the first two successful transpolar voyages, A. Vartanian, proudly announced that the next transpolar flight would be made in a four-engine bomber: "We have ended our experimental flights in single-engine ships. Now comes the real test. Under Sigismund Levanevsky, Russia's chief pilot in the Siberian region, we will send a larger craft—a bomber—over the North Pole."[4]

THE NEW MACHINE

The long distance bomber-academy (DB-A) had been designed in 1934 by the Soviet Air Force Academy.[5] Changes from the prototype ANT-6 (TB-3) included a smooth covering over the fuselage, wings, and tail rather than the earlier corrugated style; a streamlined cowling on the engines and radiators; a covered cabin for crew and gunners; a lighted cockpit; and retractable wheels that stowed inside large fairings.

The DB-A's four 970 horsepower, twelve-cylinder, Mikulin AM-34-FRN, liquid-cooled engines sent almost four thousand horsepower to four metal, three-blade, controllable pitch propellers, making the aircraft capable of speeds up to 190 miles per hour. The fuel tanks were set between the spars, leaving a wide passageway for a mechanic to reach each engine while in flight. Its heavy, multispar wings stretched almost 130 feet. The cabin could accommodate twenty-five passengers plus freight. Loaded for flight, the behemoth weighed in at thirty-four tons. The pilot's seat was located in a glazed cockpit enclosure, giving wide visibility. The only thing it lacked was cabin heat, but it was expected that the unique red wings and dark-blue fuselage (painted at Levanevsky's request to supposedly complement his family's coat of arms) would absorb the sun's rays and heat the interior through solar power.

The DB-A was designed and built in one year and, when test-flown, won four world records for load-carrying ability and speed. But there were serious criticisms of the aircraft. "There was not a flight on this aircraft that something did not break down: too much was new, too little testing," stated aviation engineer Pavel Kolesnikov.[6] Georgi Baidukov also noted that the test pilot, Kastanayev, had no experience flying "blind" in storm fronts and dense cloud conditions, a significant shortcoming in the test flying. Further, the aircraft lacked longitudinal stability, making it difficult to maintain a level and steady flight attitude. Basically, it was a tough airplane to control, and it took all one's strength to manage the rudder and ailerons and to apply pressure on the control column and pedals. "The working of the engines did not cause joy," Baidukov writes. "They often ask me, why did Levanevsky fly in an unfinished machine? Why not postpone the flight? But Levanevsky could not act in any other way—he believed in his skill. The flight over the Pole was the purpose of his life, if you like, it was the summit to which he always aspired."[7]

The DB-A was given the state registration number of N-209, and while it was being completed, the crew began rigorous training. This included working with oxygen apparatus, as they'd be flying at high altitudes in an unpressurized cabin. Three of the six crew members could operate a radio set, and the aircraft was equipped with main and reserve stations, as well as a small emergency station that could be hand-cranked. Communication equipment consisted of a radio-beacon receiver, a radio-compass, gyromagnetic and magnetic compasses, a gyroscopic semicompass, and a sun (astro) compass.

Sigismund Levanevsky was pilot and crew commander. The co-pilot was Nikolai G. Kastanayev, who had flown N-209 on its trial flights. Other crew members were navigator Viktor I. Levchenko, flight engineers Grigory

right: The pilot of N-209. Sigismund Levanevsky at Sevastopol, May 1937. (Courtesy Mike Hewitt)

below: The crew of N-209 (L to R): radio operator Nikolai Ya. Galkovsky; co-pilot Nikolai G. Kastanayev; pilot and crew commander Sigismund Levanevsky; flight engineers Grigory T. Pobezhimov and Nikolai N. Godovikov; and navigator Viktor I. Levchenko. (Courtesy Mike Hewitt)

T. Pobezhimov and Nikolai N. Godovikov, and radio operator Nikolai Ya. Galkovsky.

As flight preparations were underway, weather reports from the Arctic told of fog and rain mixed with snow, and even reports of a cyclone. Morale was not good. "We raised objections against the flight, but they would not listen to us and insisted," wrote Georgi Baidukov, who recalls that before climbing into the cockpit, Levchenko said, "Farewell, Georgi." Baidukov explained that it was not "the usual 'Until we meet,' but, 'Farewell for good. If you could be with us then there would be hope.'"[8] As for Kastanayev, his parting comment was, "You know, if I am honest— there is no confidence whatever. We have had little training." He also felt that "sooner or later the construction flaws, which for the time being were slumbering, would come to life, so it is no wonder at the depressed mood of the men before take-off."[9] Flight Engineer Godovikov embraced Baidukov and kissed him on the cheek. "Farewell, Georgi. We probably won't see one another any more."[10]

Fifty years later, Baidukov recalled that moment in a talk he gave to the Moscow Scientist Club. "I approach the lads—they hide their eyes, say goodbye with such a look, as if we will not meet anymore. But Sigismund acts the king, in his white shirt with necktie, laughing, firm handshake, embracing the soul. Well, they took off and everyone breathed with relief because they were frightfully overloaded."[11] The send-off was a mix of excitement and trepidation. A photograph of the crew in front of the aircraft shows Levanevsky smiling impatiently as he scans the sky.

THE FLIGHT

Aircraft N-209 had been loaded with four hundred pounds of food, considered sufficient for ninety days, as well as the post and furs. The full load of eighteen tons of fuel was considered sufficient for the 4,350-mile journey, expected to take just over forty-two hours in the worst weather conditions. By flying above the clouds at a height of twenty-one thousand feet along the 148 meridian, they planned to reach Fairbanks, Alaska, after twenty-five to thirty flying hours. There, they'd refuel and continue on to New York.

Meanwhile, American newspapers were receiving mixed messages. The Soviet Embassy official stationed in America, A. Vartanian, was quoted in the *Anchorage Daily Times*, on August 13, 1937, as saying that the fliers "might switch their course east of the coastal mountain range to get better weather." *The Nome Daily Nugget* corroborated this report, adding that Canadian Airways pilot Don Dawson would fly the chartered aircraft to transport Mr. Vartanian from Vancouver to Edmonton "to await the arrival of the Soviet Polar plane, where it is expected to gas up." The Soviet Embassy in Washington denied this claim, stating that no arrangements had been made for a landing other than at Fairbanks, where the fliers would receive further instructions.

Canada nonetheless began preparations. Officials in Edmonton, on learning that the Russian aircraft had a wingspan of 130 feet, sent crews to chop back brush bordering the one-hundred-foot-wide concrete "strip" on Portage Avenue (now Kingsway). Strings of lights were mounted to highlight the airstrip, and ten tons of fuel were made available. Canadian customs officials arrived in Edmonton to make regulatory inspections, and local preparations were made to control the excited crowd.[12]

Radio operators in Aklavik, Dawson City, and Fort Norman were requested to listen closely for radio messages from the Russian plane, and stand by to supply weather information. Although farther east than Levanevsky's flight plan, these sites might have been able to pick up a radio distress signal. Aklavik could also have been a destination for a pilot whose aircraft might encounter misadventure near that area; it had an airstrip, fuel, and lodging, and every Northern pilot, including Levanevsky, knew about it.

THE MESSAGES

At 23:50 hours (Moscow time) on Thursday, August 12, a message was received at flight headquarters in Moscow, relayed by radio operator Ernst Krenkel and his crew from Galkovsky's transmitter on N-209. The aircraft was flying at an altitude of 8,300 feet as it passed by Morzhovets Island in the White Sea. Three hours later, after ten hours of flight, it had crossed the 72 parallel on its way to the Barents Sea, to Franz Josef Land (Rainer Island) and on to the North Pole.

With each hour of the flight, however, its ground speed seemed to fall. Galkovsky reported that the crew was now wearing oxygen masks; they were freezing cold and did not want to move or speak. Two-way contact was lost after fourteen hours and thirty-two minutes.

At 13:40 on Friday, August 13, after nineteen hours and twenty-five minutes of flight, N-209 reported, "Flying over North Pole. Reaching it was difficult. From the center of the Barents Sea we have been flying through unbroken heavy cloud up to a height of 6,000 metres [20,000 feet]. The temperature is minus 35° Centigrade. The windows of the cabin are covered with hoarfrost. Head winds at times 100 km [60 miles] per hour."[13] At 14:32 on August 13, a further radiogram was transmitted from the aircraft. The extreme right engine was out of action due to a clogged oil feed. They were flying at an altitude of fifteen thousand feet in thick cloud, had passed the North Pole, and were reportedly two hundred miles on the Alaskan side.

Could N-209 fly on three engines? Levanevsky had thought so, for he'd stated to *Pravda* reporters the day before the flight, "One of the positive characteristics of the aircraft lies in the fact that with a flight weight of 25 tons it can fly on the two extreme engines." By this time, N-209 had used up several tons of fuel and was lightened, but three engines would use more fuel than four, a result of overloading them at the maximum possible height. The crew, numbed with cold in the unheated cockpit, would be forced to gain altitude nearly to the limit. Their furs and oxygen masks would help somewhat, but conditions could only worsen as the three engines battled on through the Arctic skies.

Ernst Krenkel spent forty hours without sleep while listening for any further messages from N-209. "Toward the end I listened standing up in order not to fall asleep," he recalled. "My ears were sore from the long hours in earphones . . . By the morning of August 14 the entire Arctic was listening for Levanevsky, as well as Soviet Polar Stations and US stations. All radio stations were mobilized, including military and amateur stations."

On August 14, a troubling report was published in *Pravda*. The government commission that had authorized the flight revealed what has been called Telegram or Message 19, the enigmatic and much discussed, numerically coded message, "48340092." The code number "48" warned of the possibility of a forced landing; speculation then centered on "3400." Did it mean "motor 34" with "00" indicating the time of landing? Or did "34" stand for the conventional assignation of an area (square) chosen for landing? The "92" indicated Levanevsky's call sign.[14]

N-209 was given orders to descend to 6,500 feet, where the crew might look for a place to land if necessary. Radio Operator Galkovsky did not reply. Later that day at 15:58, Yakutsk radio station at Cape Schmidt heard, "Everything is in order. Audibility is poor," and at 17:53, "How do you hear me? Stand by . . ."

After nothing further for nearly twenty-four hours, on August 14 at 17:44 (Moscow time) the US Army Signal Corps in Anchorage received a message. It was in number code interspersed with several Russian words, mostly unreadable, and the portion received did not give a clue as to whether the aircraft was aloft or down in the icy Arctic: "No bearings . . . having trouble with . . . wave band," the dots indicating unreadable portions. The weak signals and partial messages would indicate that at least the aircraft hadn't crashed and their radio was working.[15]

Just after N-209 had passed the pole, a further message was apparently heard at flight headquarters. It was reported secretly to Georgi Baidukov by a friend of Valery Chkalov and navigator Alexander Beliakov, and never mentioned again. "It is impossible to work in the front cockpit. We are moving back. Levchenko, Galkovsky." It remains a mystery message, seldom spoken of, never understood, neither confirmed nor denied.[16] Perhaps the cellulose (Plexiglas) nose in the navigator's and radio operator's station in the aircraft had broken, either from structural problems or ice from the propellers. "General Baidukov was very emphatic that this had happened," reports Alaska pilot Ron Sheardown, who has been involved for many years with the search for N-209. "And he probably knew more than anyone what might have happened."

It is not known what de-icing equipment might have been provided for the propellers. "There is no indication of such in any of the photos and probably no alcohol as well," Sheardown states. "They show four Venturi tubes (used to run the artificial horizon, turn-and-bank indicator, and directional gyros on the aircraft), which would have no anti-icing, as well as almost guaranteed no heat on the Pitot Tube that measured air speed. Normally the alcohol tank is in the nacelle behind the engine—with no way to fill it in flight."[17]

It is not unusual for ice accumulation to be flung from the propellers into the fuselage sides on multi-engine aircraft, causing fuselage skin damage. The ice could also have taken out the high frequency radio by breaking off the antenna. The men might have crossed into the rear cockpit where a backup navigator station was located, with a less powerful HF radio and other equipment, including a sun (astro) compass and sextant.

The Venturis icing over would take out the gyros as well. If this were the case, it is possible that the aircraft simply "spun in." It is a well-known phenomenon that pilots, without visual reference to the ground or functioning gyro instruments that replicate the horizon, will spin down and crash. Flying "by the seat of the pants" is not good enough.[18]

"In the difficult and risky career of an aviator there is one factor in addition to superb physique, technical knowledge and professional skill which determines the limits a pilot can reach," notes Ernst Krenkel in his autobiography *RAEM Is My Call Sign*. "This factor is even mentioned in solid tomes on aviation, and is called *the element of luck*. As well as willpower, talent and mastery of the art of flying, one needs a little bit of luck and this was where Levanevsky missed out."

Polar Projections Showing the Levanevsky Flight and Search Locations

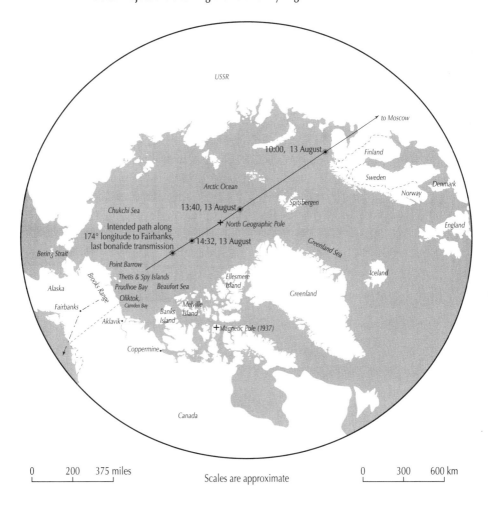

THE SEARCHES—1937

Two days after the disappearance was announced, Canadian pilot Robert (Bob) C. Randall was chartered by the Russian government to fly a Mackenzie Air Services' Fairchild 82 north from Edmonton to search the area he knew best. Randall had previously won an honorary membership in the National Geographic Society for pilot services in Yukon expeditions, and was now dispatched from Herschel Island to become the first Canadian pilot to participate in the search.

Landing at various points along the Alaskan coast, Randall spoke with local residents about the disappearance of this big Russian aircraft. At Barter Island, an Inuit settlement midway between Point Barrow on the north Alaska coast and Aklavik at the mouth of the Mackenzie Delta near the Beaufort Sea, Randall received the first indication that someone on the ground might have seen or heard an aircraft. Later, some people butchering reindeer for their fall meat supply reported that on August 13 they heard a roar like a boat's outboard motor in the fog-shrouded water. No boat had appeared, and so it could have been an aircraft.

Two Soviet pilots, Aleksey Gratsiansky and Vasily Zadkov, hurried up to Point Barrow to join Bob Randall to verify the stories. Weathering storms, they searched out from Alaska over the Beaufort Sea in a Dornier Wal (SSSR N-2) two-engined hydroplane (flying boat). Following four short flights, they were forced down by dense fog on the frozen ocean five hundred miles north of their icebreaker, *Krasin*. When Zadkov finally returned and moored near the *Krasin*, ice closed in and crushed his aircraft.

On August 14 three American aircraft joined the search to the north, northeast, and northwest. The next day, Sir George Hubert Wilkins, an Australian explorer, pilot, and navigator, as well as an internationally known reporter and aerial photographer, was called to assist. The crew consisted of Canadian pilots, Herbert Hollick-Kenyon and Silas Alward "Al" Cheesman, Australian air engineer Gerald D. Brown, radio operator Raymond E.

Booth, and navigator Wilkins. Their aircraft was a twin-engined Consolidated Catalina Flying Boat called *Guba* (an early version of the PBY-5 Catalina, a World War II patrol flying boat), which boasted long-range capabilities and modern navigation equipment.

Guba's first stop on August 22 was Coppermine, NWT, ninety miles north of the Arctic Circle on the mainland Arctic coast. The west boundary of their search area would be 82°N, 147°W, working from Rudolfo Island on the Siberian side and Point Barrow on the Alaskan side of the Arctic Ocean, with the Soviet searchers covering other areas. As the *Guba* crew flew over the Banks and Melville Islands, and northwest from Prince Patrick Island, visibility was under a half-mile. Snow and dense fog covered a panorama of ice floes, open water, and occasional land surfaces. Nevertheless, radio calls were made to N-209 every thirty minutes.

Their first flight over the ice took thirteen and one-half hours, with an average speed of 135 miles per hour. Wilkins kept vigil in the navigator's cockpit in the nose of the aircraft, using powerful binoculars that revealed broken patches of pack ice through holes in the fog. Even if the lost N-209 was somewhere below them, it would have been almost impossible to spot in such conditions. It was painfully obvious that the wheeled aircraft N-209 would have had little chance of survival, whether it landed in water or on a drifting ice island riddled with icy-peaked hummocks and water-filled holes. Mounting cloudbanks forced *Guba* to return to Coppermine.

The next day *Guba* set out again, landing in open water near Cape Russell on Melville Island to transfer some fuel from the drums to the wing tanks. After a fifteen-hour flight, they came down at Walker Inlet on the southwest side of Prince Patrick Island. "At first I was asked to search only as far as 82°N, and longitude 147°W, because Soviet fliers, working from Rudolph Island on the Siberian side and from Point Barrow on the west were to search other areas," Wilkins reported. "Later I was asked to search as far north as 88°N and between longitude 90° and 153°W."[19] Levanevsky was thought to have disappeared one thousand miles from where they were now, but maybe something would be spotted.

Guba flew back to Coppermine in cloud using instrument flight rules, the compass needles swinging wildly from the proximity to the magnetic North Pole and the vibrations of the aircraft, each reading up to forty degrees different from the last. Meteorological bureaus from Canada, the United States, Greenland, and Russia (especially from its North Pole station) sent thirty-six-hour Arctic weather forecasts.

Flight number three again met with bad weather, including "frost smoke" emanating from open water between the ice floes. Gasoline was burned so quickly that the Coppermine depot, made available courtesy of the Royal Canadian Air Force, became tapped out. The searchers moved their base to Aklavik, six hundred miles to the west on the Mackenzie Delta, and then to Barter Island, even further west and in Alaskan territory, where Captain Randall had reported Aboriginal people hearing an engine.

Autumn arrived and ice began to form on the inlets. Ice and sleet piled up on the aircraft's wings and fuselage. *Guba*'s hull became encrusted with ice, cockpit windows frosted over, requiring that they be left open or continually drenched with alcohol, and the crew wore reindeer-skin clothing to keep from freezing. On September 17 they made a last search out from Barter Island "taking off in a howling gale, snow and sleet obscuring all vision at a distance greater than 200 yards" and back toward Aklavik.[20]

In all, Wilkins and his crew made five search trips in *Guba* during August and September. They returned to New York after flying over thirteen thousand miles in a thirty-day period, twenty-seven of those days with all crew inside the airplane. "The search was ended due to two facts," Wilkins said at a gathering in New York on their return, which included pilot Herbert Hollick-Kenyon and Arctic explorer Vilhjalmur Stefansson, who had been coordinating the international search. "Open water was no longer available for seaplane use and ... 70% of all flights were done in foggy weather." They had not gone farther than 87°N because, when N-209 had passed over the Soviet camp at the pole, the radio beam indicated it went two hundred miles beyond, defining the area of search. "It is possible that a new expedition will be equipped and sent up there in October, when it will be possible to conduct the search by moonlight. Also, the new plane will have skis," Wilkins reported.[21]

By October 1937, only aircraft specially equipped to fly in the long, dark polar nights could continue, with others vowing to return in the spring. Mikhail Vodopyanov took off from Rudolfo Island in a search plane on October 7, gliding above the ice in darkness in the twenty-five-ton ANT-6, guided by compass and radio beacon. His pioneering ten-hour flight was the first search flight ever made in the dusky polar light.[22] On November 2 the sun was seen for the last time that year from Point Barrow. By November 22, 1937, Sir Hubert Wilkins and crew were heading north once more. This time he left from Edmonton in a ski-equipped Lockheed Electra, accompanied by Canadian Airways engineer Allan T. L. Dyne and Canadian Marconi Company radio engineer W. R. Wilson.

Fliers from everywhere, invited by the Soviets, had now joined the search, with generous support offered by Canada, the United States, Norway, Sweden, and Denmark, and by the Explorers Club of New York (of which Vilhjalmur Stefansson was president). Canadian companies such as Canadian Airways Ltd. and Mackenzie Air Services Ltd., as well as numerous private individuals, contributed manpower, machines, fuel, or funds.

Two long-wave direction-finding radios were set up (one across from Aklavik and the other six hundred miles distant at Point Barrow), capable of receiving and sending messages by short-wave. The Wilkins crew spent Christmas at Point Barrow, with continual snowfall and almost total darkness except during the full moon. A multitude of known weather conditions could now be used to calculate where the downed aircraft might be, but bad weather prohibited any further searches.

THE SEARCHES—1938

Wilkins and crew returned to Aklavik after Christmas and on January 16, 1938, were in the air again, in temperatures that hovered at -42°C. Ice pressure ridges starkly outlined the water's edge in the moonlight, viewed from their flight altitude of two thousand to four thousand feet. In all, four more flights were made between January and March, as far as 87.45°N along the 105 meridian, spending 284 hours 35 minutes in the air to cover 44,000 miles, of which 34,000 were flown north of the Arctic Circle.

In March 1938, Ambassador Troyanovsky of the Soviet Union announced that the Soviets were calling off all searches from the coastlines of Alaska and Canada. En route to New York, Wilkins's bedraggled Lockheed received service to its engines and airframe in Edmonton by Mackenzie Air Service's maintenance facility, giving this company valuable experience in coping with the difficulties of flying across the uncharted and inhospitable Arctic region of northern Canada. And Edmonton's citizens, so close to witnessing a transpolar pioneering flight, became conscious once again of the significance of their airport in aviation history.

Later that year, Dr. Homer Flint Kellems, from Arkansas, USA, led an expedition to Alaska in memory of an earlier aviation disaster. In 1935, American high-altitude pilot Wiley Post—who had become the first person to make a solo round-the-world flight on July 15–22, 1933, in the Lockheed *Winnie Mae*— along with American stage and screen star Will Rogers, had attempted a flight to the Orient via the northern route when their airplane crashed on takeoff at Point Barrow, Alaska.

While in Alaska, Kellems investigated a report filed by the US Signal Corps Radio Station at Point Barrow: in August of 1937, three Aboriginal people living at Oliktok, between Point Barrow and Aklavik, had seen and heard what could have been an aircraft in distress. He decided to visit the site. There, six people repeated to Dr. Kellems that on August 15, 1937, they'd been sealing and fishing at Oliktok when they heard a sound "like an

Evinrude engine" in the open sea (which significantly differed from Randall's report that the sound was heard on August 13). Two of the men, Foster Panigeo and Roger Cloud Kashak, had used binoculars to view an object moving at great speed near Thetis Island, six miles out from shore and heading toward Spy Island (six miles northeast of Thetis, and part of the Jones group of islands). Then had come a sound "like wheels splashing in the waves, passing the west end of Thetis Island." Halfway between the two islands the object had made a huge splash in the water and disappeared. The next day when the seas calmed, some men went out in their *comiaks* (skin boats) and saw oily patches in the water by Thetis Island. They observed the patches for several days, but when nothing more was seen they reported the incident to Jack Smith, a white trader at Beechy Point.

Now as Kellems and his crew stood at the point of land where the Inuit had viewed the mysterious object, he became convinced there was a connection between this incident and Levanevsky's disappearance. With the use of a compass and a dragline to find any metal objects, they checked the shallow, fathom-deep (six-foot) water of the lagoon between the mainland and the outlying islands, and then started dragging at thirty-foot depths in hundred-yard swaths near Thetis Island. Levanevsky's aircraft had stood sixteen feet high; if it had ended up on its side, a wing tip or some other part would surely protrude high enough to be snagged by the grappling hook, or at least the metal engines would trigger their compass needle.

They also went ashore to tramp over Thetis and Spy islands. Both islands were a mess of debris and pocked by water-filled holes gouged by ice and wave action—impossible landing places. A break in the beach indicated the low place where the Inuit men had seen the oil washing up, but nothing was found. "We couldn't have examined that [ocean] bottom more thoroughly if we'd had it out on dry land and raked every inch of it down with garden rakes!" Kellems stated in his report. "Yet, we found nothing. To say we were disappointed is putting it mildly."[23] Winter weather came, and Kellems and his crew abandoned their searches of "the little sand islands to the north of the dreary Alaskan coast."

On June 22, 1941, Germany invaded Russia, which put an end to scientific exploration in the north. After the war, new aircraft and helicopters took over from the old-timers, and flights over the North Pole became the norm, rather than daring adventures that fascinated the world.

THE SEARCHES RESUME

Interest in the lost aircraft waned in the postwar years, but in the 1970s and 1980s the mystery of the missing Russian bomber again begged to be solved. Areas related to the disappearance were scrutinized, including over fifty islands of pack ice that continually drift throughout the North Arctic Ocean. Their surfaces, pitted by channels of small rivers and frozen lakes, can appear similar to the tundra in winter when covered with snow. Could Levanevsky have mistaken a floating ice pack for land? The floating "island" would be carried by currents, but after some time the weight and heat of the aircraft would cause it to sink through its self-made hole. Perhaps N-209 lay at the bottom of the sea. But then again, perhaps Levanevsky *did* land the big bomber on terra firma.

In October 1967, Scandinavian Airlines' chief navigator, Einar Sverre Pedersen, crash-landed his Piper Apache on a flight from Fairbanks to Norway. When he explored his campsite, the Sam Lake area near Old Crow in Canada's Yukon Territory seventy miles inside the Arctic Circle, he spotted a log cache surrounded by odd-sized, foreign-made metal fuel barrels. Beneath the logs lay a rough casket on which rested a human skull. Could these objects be from the Levanevsky crash? The burial site at Old Crow lay on the path Levanevsky would have taken toward Aklavik. N-209 might be lying deep in the cold and murky waters of Sam Lake. But there were no barrels of petrol aboard N-209—so much for that theory. In the end, it was proven that a French trapping party had constructed the winter camp, not

Levanevsky and his crew. The search at Old Crow was, as at other places, inconclusive. In 1977, a Russian helicopter pilot reported to the director of the Arctic and Antarctic Institute that back in 1965 he'd been told that a board had been found lying on the shore of Sebyan-Kyuel Lake, in the Verkhoyansk mountain range 250 miles north of Yakutsk. It bore a partly-visible inscription: "Here perished . . . 13 August 1937 . . . N" (the letter thought to be an aircraft number) . . . and the surnames of people, one name ending in "sky." But the board was no longer available for study, having been taken by the crew of a helicopter that had crashed and burned on returning to base.

By the time investigative visits were made in 1979, no one could be certain exactly where the board had been found. Nor could reports be verified of local inhabitants having seen the bodies of two dead Russians lying on the shore with a map case, or Yakutsk hunters seeing oil stains floating on the surface of the five-hundred-foot-deep lake. However, there were no wooden boxes on board N-209. Its commercial cargo had been packed in light cloth bags or in sealed jars. Although every pound of superfluous weight had been removed in order to take on more fuel in the subsidiary fuel tank, could N-209's fuel capacity have even allowed it to reach Sebyan-Kyuel? Plus the visible wreckage of a Russian aircraft would have stood a good chance of being spotted and reported during World War II when the American Air Force had a large base at Thule, Greenland.

The next search for Levanevsky and his crew was launched by Walter Kurilchyk of Capistrano Beach, California, who turned an aviation hobby into an obsession for "chasing ghosts." Kurilchyk first became interested in searching for the resting place of an aircraft that was part of the famous American B-25 bomber mission led by General Jimmy Doolittle during World War II. He contacted the Soviet Embassy in Washington in 1983 and a deal was made: the embassy would give all assistance possible for a search for a B-25 (2242) piloted by Captain Ed York, believed to have gone missing near Vladivostok, in return for Kurilchyk's help to search for N-209. In

Kurilchyk's view, the B-25 and the Russian N-209 disappearances were not just "military anomalies." "I think that there is something particularly spooky here, any way you want to put it," Kurilchyk states.[24]

In August 1987, exactly fifty years after the disappearance of N-209, Walter Kurilchyk led an expedition on an underwater search near the village of Oliktok, between Point Barrow and Aklavik. Buoyed by Dr. Kellems's earlier search, and aeronautical surveys taken as early as 1946 that had indicated some magnetic anomalies south and west of Spy Island, permissions were obtained and a proposed grid laid out. "Neighborly assistance" was offered by Arco (Atlantic Richfield), an Alaskan oil prospecting company that was investigating the local Arctic shelf. Although ice scouring and currents would scatter remains of a downed aircraft, the large magnetometer mounted on the stern of their boat picked up about thirty anomalies—which could be pieces of *any* aircraft or even old oil drums. None were recoverable, being buried under more than five feet of sand with no dredging equipment available.

Following further research in Russia, Kurilchyk mounted another search of the area in April 1990, this one "on ice" and including Professor Eugene Konoplev from the Institute of Engineering at the University of Kiev, Ukraine, and Vaiz Yunesov, a TASS journalist. Bob Isham, an aircraft and power plant mechanic, was recruited to drive and maintain a Thiokol, an unwieldy-looking military surplus vehicle mounted on heavy tracks, used for towing a specially made, T-shaped, fifty-foot-wide aluminum sled with three mounted magnetometers. "When I left Fairbanks I had no idea what services I might render," Isham says. "I loaded my pickup with a portable generator and space heater along with tools, gasoline and oil for the Thiokol. This all came in handy when we had to change a flat tire out on the ice. There was often breakage of the aluminum sled due to the rough terrain."[25]

The trip to the site was a major undertaking—four hundred miles over the Arctic pipeline road (James Dalton Highway) from Fairbanks to

Prudhoe Bay, then forty miles west and north to Oliktok. Then came twelve- to fourteen-hour days spent in extremely cold temperatures maneuvering the Thiokol and T-bar-mounted magnetometers over and around ice pressure ridges, combing the four-square-nautical-mile grid area. Dr. David B. Stone, associated with the Interior & Arctic Alaska Aeronautical Foundation (IAAAF), rode in the back of the Thiokol, directing Bob Isham to steer the desired course, often stopping to wait for a third satellite to appear over the horizon so they could get an exact fix on their location. "Any detected anomalies were saved to disc using a laptop computer and a GPS system to mark the exact location. Definitely hi-tech," states Bob Isham. Following the field work, Dr. Stone and his assistants assimilated the data, using computers and a Mini-Ranger survey system to mark the exact locations of the anomalies. "The expedition recorded over 50 anomalies," Kurilchyk states in *Chasing Ghosts*, his chronicle of his searches for N-209, "some of which could reasonably correspond in size to the N-209 engines."[26]

Kurilchyk continued his research in collaboration with directors from the Russian Far Eastern Museum in Arsenyev, Primoriye Region, to whom he offered his full search records. He, in turn, received some "security-significant," previously classified information on the B-25, and a promise to pursue further facts on its fate.[27]

In 1990, Kurilchyk was contacted by Yuri Salnikov, a Moscow television film director (*Unknown Quadrant of Levanevsky*) and author of a book on Levanevsky titled *The Life Devoted to the Arctic*, who encouraged him to develop a team to once again search Oliktok. Later, in his article "The N-209 Enigma," Salnikov asserts, "It is our moral obligation to resolve the N-209 enigma."[28] As recently as 2002, Salnikov has encouraged Kurilchyk to assist him in developing a team to continue the search for N-209. "Yuri Salnikov has expressed an interest in making an expedition," Kurilchyk acknowledges, "but his problem is lack of official support and lack of money. If he did get serious and have the support, I would have participated."[29]

At the Arctic Circle. (L to R) Dr. David Stone, Everett Long, and Bob Isham, with Thiokol track vehicle. (Courtesy Bob Isham)

Some skeptics have denounced the Thetis–Spy Islands area as a possibility, citing the discrepancy between the dates of the Inuit hearing an engine on August 13, as reported by Canadian pilot Bob Randall, and Dr. Kellems's notation that this occurred on August 15. If it was August 13 (and if it was N-209), perhaps it had landed on some ice and taken off again; however, that seemed impossible in the difficult circumstances. If it was August 15, it could have been a search plane that crashed into the ice-strewn sea. And perhaps the Thetis–Spy Islands had no relevance at all to the missing bomber. "The question as to why I pursued the search for both of these planes is indeed a provocative and personal matter," states Walter Kurilchyk. "When I think of the B-25 I am thinking in terms of the early 1940s. My mind is back to those days . . . you have to dwell in the psyche I guess, to understand. I remember those days clearly, the men, my friends, family, etc., all have something to do with it. And it is also the matter of 'history.'"[30]

In the conclusion to his detailed and fascinating book, Kurilchyk states,

"I am convinced that the elusive two historic airplanes that have captured my thoughts, time and resources for over fifteen years have been found. Other persons may be the first to discuss them publicly, but I hope to be somewhere nearby for their final flights from obscurity of so many years."[31] "There is no contradiction in my comments," Kurilchyk later writes in an e-mail to the author. "I believe that the KGB have information regarding the disposition of the B-25 and I am assuming that the circumstantial evidence sufficiently reflects the N-209 location. Of course the 'Best Evidence' is the physical hard evidence. But when you have such substantial circumstantial evidence, you have to take a stand. I look forward to comments from some of your readers."[32]

Mike J. Hewitt of Sheffield, England, who has followed the mystery of N-209 with avid interest over the years, offers these conclusive comments:

> Whenever a plane disappears there are always reported sightings which are way off target. The Spy/Thetis sighting always looked to be one of these . . . i.e., it was a Canadian search plane which nobody was aware of. Although I am no expert, it always seemed to me that it was most unlikely that the N-209 could have flown that far with engine problems. I am inclined to agree with what Krenkel writes in his book, 'Aviators conjecture that the cloud was probably so low that it reached right down to the ice, merging with the fog. The most likely explanation is that the aeroplane flew lower and lower until it ploughed into the ice.' Failing that—was Levanevsky defecting? This theory has been discussed in Alaska but without a shred of evidence to support it . . . so I think it's another red herring. However, I doubt very much if the Russians withheld anything from Walter Kurilchyk. I cannot imagine why they would have spent all that money looking for N-209 had they known something. There is nothing I have read to suggest that General

George Baidukov was other than an honourable man. Yes, it must be a sensitive subject for Walter, but I don't think he was duped.[33]

Ronald C. Sheardown is a Canadian pilot flying out of Anchorage, Alaska, who became involved in searching for the missing Russian fliers in 1962 at the request of the RCMP when an old airplane wreck was spotted by a helicopter pilot west of Pond Inlet on Baffin Island.[34] The search was inconclusive, but Sheardown again searched the area in 1992 and 1993, intrigued by the idea of Levanevsky having followed the Great Circle Route (the shortest, most direct route between two points on Earth's surface).

In 1999, Sheardown flew to Camden Bay, seventy miles east of Prudhoe Bay, with his An-2 to locate and inspect a side sonar image. The object was seventy feet long with subsidiary shadows that could be wings, and was submerged in twenty-one feet of water, two and one-half miles offshore in the Beaufort Sea (west of the North Slope village of Kaktovik). A seven-person team returned on May 27, 2000, staging out of Deadhorse, Alaska, to operate on the floating shore-fast ice of Camden Bay. A five-foot-diameter hole was drilled into the six-foot-thick ice, into which was lowered a special underwater remote operating vehicle (ROV) with a video camera where the anomaly had been discovered. No man-made object was seen and the search was concluded on May 29.

During the Memorial Day weekend in May 2001, the US government's Minerals Management Service, the University of Alaska Fairbanks, and Ron Sheardown, who supplied an aircraft and transportation, conducted another underwater search at Camden Bay. They again drilled a hole through the ice and dropped down a battery-powered ROV with a remote video camera that could travel up to 650 feet underwater in any direction. Nothing was found. Sheardown concluded they might have been searching three hundred to four hundred feet too far north, and hoped to repeat the search.

In June 2002, an eight-member crew from the University of Alaska

Fairbanks, Geophysical Institute Electronics Shop, and Sheardown made a further search for Levanevsky's aircraft. They again drilled a hole in the Arctic Sea ice at Camden Bay, and lowered the ROV to view a "60-foot cigar-shaped object" that had been discovered during an oil exploration survey. Though the object proved not to be the wreckage, the search renewed interest in the sixty-three-year-old mystery of the aviator's disappearance.

Sheardown has not given up the search. "My philosophy has been to eliminate theories and possible areas. This is what we did on Baffin Island and Camden Bay. My theory base has been [that Levanevsky] turned left to the Arctic islands of Canada after losing an engine on N-209, 300 kilometres (186 miles) after crossing the North Pole. N-209 had three ventures on the left side of the aircraft. If they iced up they would have no gyro operating instruments and could have lost control in cloud. I have stated this for many years, as a pilot who has flown the Polar routes many times." Sheardown vows, "We will find Levanevsky's aircraft and if not us, then our successors."

The Brooks mountain range in northern Alaska is another area that has been considered a possible search target, although its valleys remain bare of snow until the end of August and sometimes well into September, and there are few glaciers that could hide such a massive aircraft. Also, a large number of pilots fly over the Brooks range, and it seems unlikely that N-209 could have crashed there and not been discovered. "There are 10,000 licensed pilots in Alaska and 6,000 private aircraft," notes Ron Sheardown. "Thousands of hours are flown each year in the Brooks Range by airplanes and helicopters. Most of the area has been geologically and geophysically mapped over the last 100 years, mostly in the last 50 years." But navigator Valentin Akkuratov feels it is important that any and all areas be considered. "Any version, even the most imaginative, may be feasible," he states. "It is difficult for a person unfamiliar with the Arctic to imagine the whole complexity of a search in polar conditions. As my friend Anakul, an Eskimo from Wrangel Island, said, 'There are aircraft, there is bad weather, there is good weather, there are no aircraft.'"

Other theories point to a crash site far from Alaska. Continued analyses of Levanevsky's last radio messages produce reasons why the Canadian archipelago should be thoroughly searched, especially Ellesmere Island, a remote and uninhabited region topped by ice mountains. It might have been possible for Levanevsky to make a forced landing from the ocean side of Ellesmere Island, which is indented with fjords, such as the southern shore at McClintock Inlet. This area is three to four miles wide and about thirty miles long, and to the south rise the highest mountains on the island, covered with glaciers, as described in one of Levanevsky's last messages.

Ghost ships embedded in ice have been the subject of many legends. An American Douglas transport aircraft, found abandoned on the drifting ice in 1952, had traveled about in the Arctic Ocean for two years. Perhaps the N-209, encased in its own icy coffin, is still "sailing" the seas, opening the possibility that it could have reached the northern extremity of Ellesmere Island.

After studying past searches for N-209, Ron Sheardown has decided to concentrate on areas not thoroughly examined. He has often reviewed a message from Levanevsky received by the steamer *Batum* which was interpreted as 83°N and 179°W— but what if it was the 79 meridian instead, and the same latitude? That would place N-209 near the entrance to McClintock Inlet on Ellesmere Island. Following a trip to the North Pole on the Russian nuclear icebreaker *Yamal* out of Spitsbergen, Sheardown, along with Jeff Helmricks in his Twin Comanche, conducted a preliminary search of McClintock Inlet during August 2000.

Further explorations for N-209 at McClintock Inlet were planned for August 2002, the sixty-fifth anniversary of Levanevsky's disappearance, by Sheardown, along with the Moscow Regional Center of the Russian Geographic Society—"Expedition Center-Arctic" or EC-ARCTIC—founded by the Russian Geographic Society and Russian Polar Explorer Association. Although the 2002 expedition did not happen, and funding shortages have impeded further intensive searches, the mystery of N-209 has been neither concluded nor dropped. As Magadan-based journalist Mikhail Ilves notes, "Although the Arctic was and still is the most aggressive protector of such mysteries, the researchers do not give up hope."[35]

TRUE OR FALSE?

Russia's Shortcut to Fame, by Robert J. Morrison of Vancouver, Washington, published in 1987, suggests that literally all the Soviet polar flights of the 1930s were hoaxes. Morrison's view is that Josef Stalin wanted Germany and Japan to believe that Russia had long-range bomber capabilities, "to hype Soviet air prowess and purge enemies," and that the Russian government had "fooled the world for more than 50 years."[36] Why would the United States go along with the "hoax"? Morrison opines it could have been an attempt to please a potential ally (Russia) against Hitler's Germany.

Another hotly contested theory is advanced by former *Pravda* journalist Oscar Kurganov, a long-time associate of the Levanevsky family and personal friend of Sigismund. Kurganov, who was initially invited by Levanevsky to accompany the flight, postulates that publicity for the famous over-the-pole flights conveniently covered up "Stalin's secret plan" of ongoing trials of a number of dissidents, whose punishment was death. He further recalls the disgrace Stalin felt over Levanevsky's earlier flights from America to Russia being made in an American Vultee rather than a Russian aircraft.

Politics were certainly at the forefront of Levanevsky's career. Stalin initially wanted him to postpone the famous flight of 1937. When permission was suddenly given, Levanevsky commented to Kurganov, with an understandable sense of foreboding, "Apparently one more trial is about to take place." In Kurganov's article, published on August 24, 1996, in *Izvestiya*, titled "Stand by . . . The Mystery of the Death of Flier Levanevsky," he asserts that there was indeed a "mystery package" included with the cargo on N-209.

"Two hours before the start a young man came to Levanevsky at Shelkovo airfield," Kurganov writes. "He produced his attestation, signed by Yezhov himself, and delivered a cardboard box which was packed and sealed. 'A gift from Stalin to Roosevelt's wife—two fur coats and a jar of black caviar.' Levchenko took the box and carried it into the aircraft's cabin. Saying good-bye to me, Levanevsky said quietly, 'There is a mysterious box flying instead of you.'"

"According to the thoughts of the author," journalist Aleksey A. Burykin notes regarding Kurganov's contentious article, "in the box instead of a gift might be a bomb placed with a clock mechanism, which was acti-vated soon after the aircraft crossed over the Pole. Just this, in the opinion of O. Kurganov, explains the lack of wireless messages about a possible crash and forced landing." Burykin adds that, in truth, there were only fragmen-tary messages and indecipherable radio signals at an interval of more than a month from the time of the aircraft's disappearance. "This means that either the given account is incorrect, the explosive device did not destroy but only damaged the aircraft, or reports about radio signals being received in very different places from Archangel to the Sea of Okhotsk are sheer misinfor-mation." While such an account has "the right to be in existence," Burykin acknowledges that it is very difficult to substantiate.

Was it caviar and a fur coat for Mrs. Roosevelt, or was it a bomb? As predicted by distinguished USSR navigator Valentin Akkuratov, "When there are not enough facts, hypotheses usually appear."

AN EVERLASTING ENIGMA

There is some speculation that Levanevsky, after initially failing to reach the North Pole, simply couldn't face a second failure. When he experienced dif-ficulty he might have chosen death over dishonor and plunged his aircraft into the sea. Others theorize that the flight crew might have staged their own disappearance to seek protection in the West, far from the "repressive Soviet monster."

"The Russians are very good at taking in all the information and giving very little back," writes Ron Sheardown. "I am still convinced that N-209 could have flown two different routes, and may be the reason that it has not been found." If N-209 had followed the Great Circle Route to Oakland, it would have flown over the top of Baffin Island and over Edmonton. Sheardown asks, "Why did Varatanin, a KGB officer with AMTORG, the Russian trading company in New York which was well known as a front for the KGB, want to go to Edmonton from Seattle via Vancouver the day the aircraft did not show up? I tried to meet with him when I met with General Baidukov but they said his health was not good. He was 91 in 1993 and has since died, as has Baidukov." When Chakolov was interviewed in June 1937 in New York, he stated that the only problems they had were over Baffin Island—again the Great Circle Route. "I doubt anyone will ever know the real answer on this question," Sheardown says. "This is why I tried to get the KGB, foreign affairs and Northern Sea Route files—with a little success—but I am convinced there is much more in those files."

Sheardown's guess is that they probably landed or crashed in the Great Circle Route. "The forecast winds for that day on Levanesky's wind map—which I have a copy of—would have been from the southwest at 20 to 40 knots. There was a low pressure system northwest of Tiksi and north of Khatanga on the Russian side, making it highly unlikely they would have tried to head for the north coast of Russia, as in their radio message at the North Pole they reported 100 kilometres (60 knot) headwinds."

If they went down on the ice, no one will ever know. All that can be done is to eliminate theories one at a time. "I think with the activity in Alaska, the Canadian Arctic, Russia and Greenland over the past 50 years it would be hard to hide an aircraft on land without someone seeing it," Sheardown concludes. "On the ice or glaciers would be impossible. Underwater has some hope. I do not have much hope in the geophysics that were conducted at Spy and Thetis—magnetics will get many anomalies that could be anything, probably bedrock alteration. A good survey would include underwater magnetics, side sonar and camera. The oil companies are carrying out this type of survey and someday they may stumble onto something like the Camden Bay anomaly." "Some stories have more to say than others," Ron Sheardown writes in March 2004. "It would be nice if they could be put on one page, but that is not the way it is."

This unsolved mystery continues to intrigue aviators and scientists the world over. They recall the old stories, recreate scenarios, reread every piece of printed or radioed information, and continue to raise funds to mount searches in the hope that new technology will aid in the discovery of N–209 and its legendary crew. Streets in Moscow and Kiev still bear Levanevsky's name, as does a museum in the quiet town of Sokulka, and a lonely island out in the Arctic. Numerous newspapers and magazines have chronicled this story, as well as books in various languages.

A television film by Russian producer Zbignev Kovalevsky, titled *The Fate of the Levanevsky Brothers*, was narrated by Levanevsky's sister Sofia Kornatskaya, who lived in Konstantin, Poland, in the house bequeathed to her family in 1939 through the Soviet Embassy. (Sigismund's brother, Joseph, and his friend F. Zvirko both studied aviation at the flying school in Demblensk. When the two men had their cards read to learn their fortunes, black aces appeared in their readings. Both met their death in airplane accidents shortly thereafter.)

In May 2002, Moscow aviation journalist Yuri Kaminsky, who told Levanevsky's story in his book *The Life Devoted to the Arctic*, organized a celebration in Moscow on the one hundredth anniversary of the famous pilot's birthday. It was attended by several relatives of the crew of N–209. Kaminsky also produced a thirty-minute film for Russian television about the Kremlin Flights, which includes rare footage of Levanevsky. On this side of the world, visitors flock to see the story recounted in a display at the Alaska Aviation Heritage Museum at Lake Hood.

The day before takeoff of aircraft USSR N-209, Sigismund Levanevsky wrote, "In the history of aviation there has not been the opportunity for one country in a short period to prepare and equip three long-distance, non-stop, flights. 1937 will go into the history of world aviation's outstanding never-forgotten pages."[37] In *Russian Lindberg—The Life of Valery Chkalov*, General Georgi Baidukov concludes, "As I see it, Levanevsky perished on the Arctic ice, somewhere just across the Pole. This was his fate. And although he only lived to be 35 years old, he was a flier with a capital letter."[38] That statement, too, encourages debate. Ever questions, never answers, to the disappearance of these famous Stalin's Falcons.

THE NORTH RECLAIMS ITS OWN

People who live in the North learn to respect the erratic climate and vast emptiness of the geographic area they call home. Aviation people especially see the results of lack of preparation or plain bad luck when they are sent on medivacs to pick up the pieces of a flight gone wrong, and when veteran northern flyers go missing, comrades refuse to allow their fates to remain unknown. They search and search for signs—bits of glinting metal, a dark spot on the white tundra, a white flash in a lake or amidst the green canopy of trees—anything unusual. Many stories lack a conclusion, leaving families to await news of a loved one that may never come. Yet, others are allowed closure many years later, when the lost are finally found and the remains brought home at last.

GAVIN EDKINS—A FAMILY'S HOPE, A COMMUNITY'S VIGIL

In 1996, Stan Edkins and his family—sons Kenn and Gavin, and daughters Lorell and Janice (their mother Evelyn had died in 1971)—lived in Fort

Smith, Northwest Territories, and flying was their business. Stan's company, Aeropac Flights Ltd., co-owned with Marilyn Walker, operated float-equipped aircraft, including the oldest flying Norseman in Canada, a 1938 Mark IV (CF-BFT, Serial #17), and leased several Cessnas. Gavin Edkins was employed by the Town of Fort Smith. He and his wife, Sonia Lynne Beaulieu Edkins, had two children, Shadow Lynn (four years) and Angela (whose first birthday was coming up on July 1, 1996).

Gavin had inherited a love of flying and held a private pilot's licence. He hoped to join the family business and had already completed most of the requirements for his commercial pilot's licence, including twenty hours flying Instrument Flight Rules (IFR) logged in 1995, and also some training on low visibility and low altitude flying. A 1964 Cessna 150D (CF-WYZ) had been leased from Tom Steele of Sherwood Park, Alberta, so Gavin could complete his necessary time. Now the last few hours of the required two hundred could be gained on a flight to Red Deer, Alberta, where he would meet with Dennis Cooper of Sky Wings Aviation Academy Ltd. to test for commercial certification.

The thirty-six-year-old pilot took off from Fort Smith for Fort McMurray and Red Deer on May 19, 1996, at 6:30 AM, flying the white Cessna with a red stripe along the fuselage and wing tips and a black tip on its tail. Gavin filed a direct route plan (Fort Smith–Fort McMurray–Red Deer), but said he would likely follow rivers and roads as he was a good map reader. He refueled at Fort McMurray using fuel from his jerry cans, departed there at 9:49 am, and called at 9:54 to report being on course. He was never heard from again.

The Search Begins

When the Cessna 150D did not arrive in Red Deer as scheduled, Dennis Cooper called Stan Edkins: "Where's Gavin? He's a couple of hours past his ETA!" The weather had been fair enough to give Gavin Visual Flight Rules (VFR) clearance all the way to Red Deer. Only later, after he was on his way,

above: The Edkins family in 1968 at Fort Smith, with Gateway Aviation Cessna 185. Back: Stan, Kenn, Evelyn, Lorell; front: Gavin and Janice. (Courtesy Stan Edkins)

right: Gavin and Sonia Edkins with their children, Shadow Lynn and Angela. This photograph was taken shortly before Gavin's disappearance. (Courtesy Stan Edkins)

did Air Radio in Edmonton receive two pilot's reports indicating that storms had started seventy miles south of Fort McMurray, continuing to Lac La Biche, a situation Gavin wouldn't have been aware of until he was right in it. This could have interfered with his flight plan, causing him to fly west or east, expanding the search area to 10,500 square miles.

The Red Deer Flight Service Station notified Rescue Control Centre (RCC) Trenton and the Civil Air Search and Rescue Association (CASARA). *Search & Rescue (SAR) Edkins, Case No. T96-A0250* was immediately filed, with Captain Greg Illchuk, 435 (T&R) Squadron, 17 Wing Winnipeg, as search master.[1] Search headquarters were set up at Fort McMurray under the command of Captain Neil Kinley, who assured the public that all would be done to find the northern flyer and emphasized the pilot's own knowledge of wilderness conditions: "There are certain plants that he could dig down deep for, and he should have no trouble finding water. He comes from the North, where survival is not taken for granted. He had a heavy [black leather] jacket and luckily it has not been all that cold."

The day after Gavin's disappearance, CASARA detachments from Fort McMurray, Edmonton, and Red Deer brought six aircraft, and Northwest Air Lease flew Gavin's father, Stan, to Fort McMurray to help. "Have you contacted Alberta Forestry?" Stan asked the search master. "They have the best weather reports anywhere in Canada." No, they had not. "Has your helicopter landed to check highway service stations and camps?" he asked. No, that would cut into search flying time.

"At this point I was aware of what a stupid setup I was facing," Stan later reported. He called Dale Huberdeau, chief of forestry in Fort McMurray, and requested a copy of the lightning strikes in the area. Dale agreed to check with all his tower operators. He also contacted the fish and game departments to obtain phone numbers of big game outfitters to ask if anyone had spotted the aircraft. The situation was critical. Working together was the only way to a quick and successful resolution.

That night in Lac La Biche, lightning and thunder crashed around the house of Marion and Alex Kuziemsky. At three o'clock in the morning Marion awoke, soaked in perspiration. She shook her husband awake. "I know where a missing plane is." Marion reached over for a piece of paper and drew a picture. "Here's where it is," she said, indicating a spot halfway between Lac La Biche and Conklin.

"Sure," Alex yawned and went back to sleep.

But Marion couldn't rest. She got up and went to sit by the living room window. She had seen a white and red airplane with a black tip on the tail being tossed about in a rainstorm, then an explosion, like a huge firecracker in the black sky. She had seen it, but who would believe her?

On May 21, the search boundary expanded to follow the railroad. Joining the fleet were an Air Force deHavilland Buffalo R (Rescue) 451 from British Columbia, Twin Hueys R–011 and R–023, as well as seven CASARA aircraft. Private pilots flew a Turbo Beaver and a Cessna 185 along the Athabasca River route at the request of "next of kin." Over one hundred people staffed telephones and dispatch centers around the clock.

Stan Edkins, however, discovered his presence was not particularly welcome in the SAR offices. "I was getting more information from people wandering around than from their office," Stan says. "The military was in charge and the bush pilots were in the way." Stan asked that bush pilots be allotted search areas that would not impede the military's efforts, to no avail. Even the CASARA groups had to fight for a piece of the action, and donated fuel was not being parceled out to local pilots. Finally, two pilots phoned Trenton, blasting a system that ignored local wisdom and forced many searchers to pay their own expenses.

Tension continued to rise. The military appeared to organize searches based on what an IFR pilot might do, while the locals stressed that Gavin was flying VFR, which meant he'd go under or around a cloud. From reports of the storm's severity, Stan was sure Gavin would have detoured rather than dipped, which would open up an entirely new search area.

Marion Kuziemsky again awoke at three AM, as she was destined to do throughout the search. In her dreams she saw, again and again, a white airplane with red markings, and a black tip on its tail like a weasel's, being tossed around in a rainstorm until poof! a fiery explosion and it was gone.

By May 22, searches had been completed over 95 percent of the area at the one-thousand-foot level (one nautical mile visibility—NM viz), with 15 percent covered twice at the five-hundred-foot level (one-half NM viz). Stan Edkins anxiously viewed the daily reports. Two spotters in Hercules aircraft, flying at five hundred feet and at reported speeds of 150 knots, would flash by trees as if they were blades of grass. The more he questioned their methods, the more despondent he became.

When told that the operating cost of an Air Force Hercules was seven thousand to eight thousand dollars per hour and that it transports two spotters, Stan Edkins compared that to commercial rates of $120 per hour for a Cessna: 62 Cessna 172s each carrying three observers could replace one hour of flying time for a Hercules—or 186 spotters in the air versus two in a Hercules. The three Hercules being used carried six spotters in total. He compared that to 62 Cessnas carrying 186 spotters and felt sick. Then he went even further. These 186 Cessnas could blanket the entire search area in three days, at a speed and altitude that would optimize the probability of seeing something on the ground. Compared to a Hercules, which needed a two-mile turning radius, the grids could also be much tighter.

By day five of the search, 100 percent of the area had been searched at five hundred feet and one-half NM viz, and a second coverage to 65 percent. Twelve CASARA and four private aircraft were in the air, in addition to the military's Griffon R-415, Labrador R-307, and Hercules R-310. (The Buffalo had been released that day.) A faint ELT-like signal was picked up by R-307 on 121.5 MHz using their BR-15 receiver near the hamlet of Anzac, but it was too weak to trace. Because power lines could pick up these signals, and amplify and transmit them, thus making the transmission appear to be coming from another location, the military issued an "Operation Lights Out" order in an attempt to localize the signal. This interrupted service for five hours to several small communities, as well as to oil facilities along the power line between Lac La Biche and Fort McMurray.

By the second week, the search had intensified to involve spotter training for more than two hundred people in order to crew the large number of military and CASARA aircraft. In fact, nine military and fifteen civilian aircraft from Alberta and the Northwest Territories had now clocked nine hundred flying hours and covered 7,500 square miles. The onerous lodging, transportation, and "fight feeding" requirements were managed at search headquarters using contributions by municipal governments, First Nations band councils, aviation-related companies and organizations, as well as various corporate and personal donations. This was becoming the largest and most intensive search and rescue mission ever carried out in Alberta's history.

When a week passed with no success, Captain Kinley assured people, "The search will continue until the area has been completely covered, all leads have been exhausted and all possible efforts have been made to find the plane."[2] Tantalizingly, R-307 was still picking up the 121.5 MHZ signal, so ground electronic searches along the power line had continued. However, it was concluded that this signal was not coming from the missing Cessna, and the ELT investigation ended.

Again and again the community was praised for its efforts: children donated their allowances and adults donated time, cash, and even full paychecks. The Salt River First Nation #195 near Fort Smith raised funds to send twelve spotters to Fort McMurray's search headquarters. "I am amazed and gratified at the support of Fort Smith," Stan Edkins said. The searches went on, but they weren't enough.

Marion went to bed the night of June 1, and as usual awoke at three o'clock. This time she'd seen a man who she thought to be the pilot's father. He was wearing camouflage coveralls and a cap of the same color, and was standing in an area of dried grass with his back to her. He was desperately looking for the airplane, and in the correct direction toward the east, but from the wrong vantage point. Marion watched him closely. He'd be about five feet ten inches tall and in good shape—an outdoors type of man.[3]

The man turned and plaintively asked, "How would you like it if it was your son that was lost and no one said anything?"

Marion woke up, sweating, breathing heavily. She had to tell someone! But who?

After two weeks, nearly twelve hundred flight hours had been logged by searchers covering fifty-four thousand square miles of land. The Search and Rescue operation had brought in all available military aircraft (twelve Canadian, three German), and more than twenty CASARA aircraft had been involved, as well as many private aircraft. The entire area had been searched twice at five hundred feet with one-half-mile viz, four times at fifteen hundred feet with three-mile viz, and once at one thousand feet with one-mile viz. A study of weather patterns at the time the Cessna had gone missing indicated it was most likely to be found on the flight path between Mile O and Mile 180, which had received extensive coverage.

On Sunday, June 2, both the military's Transport and Rescue unit and CASARA called it quits. "We'll fly a few more hours in the morning, by which time we'll have completely covered the search area, some parts several times," stated Captain Neil Kinley from search headquarters. This decision was unacceptable to the Edkins family. "A lot of people think because it's been so long he can't be alive, but people survive," Sonia Edkins stated to *Calgary Herald* reporter Rick Mofina. In fact, one of Gavin's oft-repeated sayings to his wife was, "Hon, we're survivors." She clung to that philosophy.

Marion could stand it no longer. Her dreams were too real. When she read in the newspaper that the search had been called off she knew she must report her visions. She walked downtown to the RCMP station. "I want to talk to you about the missing airplane," Marion announced. "I know where it is."

"You do?" asked the officer. "Well, come in then."

Hesitantly, Marion told of seeing the airplane and its markings in her dreams long before she'd read the description in the paper, and of seeing the pilot's father. She then pointed out the location on a map on the office wall, in a lake halfway between Lac La Biche and Conklin near a railroad point known as Philomena.

"We appreciate this information, but the area you mention has been searched over six times and they've now called it off," the constable said. "Also, our office has nothing to do with the search. That's the responsibility of the detachment at Fort McMurray. Or the Search and Rescue operation. Call them."

"What's their phone number?" Marion asked resignedly.

When she got home she called the Search and Rescue unit in Fort McMurray, and asked to speak to the captain in charge. Again she related her story. "What color is the pilot's father's hair?" was the first question asked of her.

Marion hesitated. In her dream she'd viewed the scene from above and the man wore a cap. "It was hard for me to see that," she stammered. "But please, make sure that he gets my message." She thanked the captain for his time, and hung up the phone.

That night the dreams returned, stronger than ever. This time the pilot himself spoke to her. "Use my three o'clock and your four-thirty," he said. "I'm in that area." He then instructed her to go to her spare bedroom dresser where in the left top drawer she'd find three rulers. "Take them and use them."

A Family's Hope

In an interview with the *Slave River Journal* on June 4 from the Fort McMurray search headquarters, Stan announced that he was determined to follow any and all leads until his boy was found. He wasn't knocking the military's efforts, he said, simply their "absolutely rigid" programming. "I've told my family that if I ever go missing, don't let Search and Rescue take over. Call the bush pilots."

Noted northern pilots Jim McAvoy and Joe McBryan offered help— gestures that brought renewed vigor to tired and disappointed crews. "Joe

McBryan phoned me at home to relieve on the [Cessna] 185 with Dan McGonigal," McAvoy said. "I had worked with Stan Edkins at Air Tindi, and I was glad to be of some help. I didn't know the kid or his experience, but if you leave McMurray to Red Deer you have a 'paved airport'—the highway—all the way. Something was wrong for him not to follow the road. I've landed on the road in the 185 myself, 40 miles south of McMurray. He likely went east to follow the railroad."

At this point, the incompatibility of methods and ideas between the military and civil crews became painfully apparent. "Stan believes, with the trained eyes of the bush pilots volunteering their services to help find Gavin, their team will be able to do a more substantial search of suspect areas," stated the *Journal*. "My big fear is that he's injured, starving, and can't get to water," Stan lamented, echoing the unspoken consternation of others who diligently kept up search efforts.

In the morning over breakfast, Marion told her husband of the latest dream. To prove its validity she went to the dresser. There in the drawer lay the rulers. She must have placed them there some time ago and forgotten about them. Now Alex was intrigued. He and Marion spread out a map on the table, and stood holding the rulers. What to do?

That afternoon she called the RCMP office and spoke to Corporal MacIntosh. "All right," he said, "come down and we'll try these rulers on a map."

They spun the rulers about, pointing them this way and that, trying to follow the pilot's instructions regarding the "three o'clock" and her "four-thirty" positions. From this they pinpointed a location: three lakes north of Lac La Biche, a large one and two smaller ones. The rulers pointed to the third.

"Please make sure this gets to the pilot's father," Marion pleaded. "And also, would you please keep this between ourselves."

"Why?"

Marion's face reddened. "Well, you see, we believe that when a person goes before his or her time, after 40 days the soul goes to Heaven and the spirit walks the earth until it finds someone to complete what they cannot complete themselves. They are what we call the 'little people,' the ones who've passed before their times and appear to us in dreams."

The corporal stood silently, paper in hand.

"I happen to be the one that Gavin Edkins picked," she said. "It's before the 40 days, but he's definitely speaking to me. I have to pass it on. I just don't want this news to get out 'on the street.' People wouldn't take it serious."

"I understand," said Cpl. MacIntosh. "Our aim is to find the pilot and the airplane. We'll get the message to Stan, the pilot's father—the one you saw in your dream."

As Marion walked home, she looked up into the sky. Airplanes flew overhead, searchers who hadn't quit. She felt better now. Someone had listened.

Marion had another dream the night of June 4. "Go and find me!" the pilot instructed. "Get your head out of the sky. Concentrate on what is happening on the ground!" He then gave her a list of items to bring along, including two whistles.

"I don't have any whistles!" Marion protested.

"Yes, you do," the pilot replied. "Check the basket on top of the fridge."

Marion awoke and looked at the clock: 2:00 AM. Curious, she wandered out to the kitchen and checked the basket on the fridge. Two whistles nestled among the odds and ends. She fell asleep on the couch and awoke an hour later to a smell of shaving lotion being carried on the breeze through the open window. She looked out. No one was there.

When her husband got up, she told him of the instructions she'd received in her dream. That morning they loaded their 4x4 vehicle with the required items and drove to Philomena, a scattering of cabins along the CN railway tracks, approximately twenty-five miles north of Lac La Biche. There they crossed the tracks and continued driving for three miles until they were stopped by a washout on the road. They got out and walked until a bridgeless river prevented further exploration. Marion took out one of the whistles and blew on it. Immediately the smell of shaving lotion drifted to her on the southeast wind. They followed a cutline east for a mile or so until they reached a slough. After returning to their vehicle, they drove around trying other roads, but were blocked by high water in the Owl River and mud holes that swamped each new trail.

A call from Cpl. MacIntosh the next day, asking if she'd discovered anything, gave Marion further encouragement. When her dreams returned that night, she asked the pilot to describe exactly where he was. He didn't know, but he was underneath something cold and he was "scared stiff." Marion believed he was still alive, but they must find him—and soon.

In her dream the next night, Marion Kuziemsky was miserable and told the pilot so. She felt physically sick, and was tired of being the medium in this mystery.

"My name is Gavin Edkins," the pilot said quietly.

"No!" Marion cried, and woke up. But when she returned to sleep the dream continued.

"Put a marker on the map!" the pilot ordered. "It will show you where I am, within a five-mile radius. Then I'll leave you alone."

In the morning, Marion and Alex again packed up their vehicle and went on a search, parking at Philomena and walking down the CN railway tracks toward Conklin.

At Mile 154 they stopped. Something attracted her attention from the hill to the east. The smell of shaving lotion again permeated the air. Leaving dandelions on the tracks as markers, they continued walking to two hundred yards past Mile 156—the exact viewpoint she'd seen in her dream. She hung a ribbon on a sign on the tracks, then she and Alex slowly turned back. It was enough for today.

On Saturday, June 8, the dream returned. The pilot spoke more politely and patiently now. "I know that I made you take that long walk and place the marker on the tracks, but still no one came for me."

"Give me more time!" Marion cried, and awoke. She felt haunted, afraid to go to bed. She'd close her eyes and the pilot would appear.

"Come back!" he'd cry. "Come back!"

"No! Your messages are too vague, and they never end!"

He became impatient. "You swim like an otter! What is keeping you?"

Could this mean he was under water? Dear Jesus, please help me.

The search headquarters, now staffed by just family and friends, received continuous tips, as well as local funding. Radio and television stations joined with print media to broadcast updates and appeals for help. Local CBC reporter, Patti Kay Hamilton, worked tirelessly as both a spotter and reporter.

Dreamers and psychics reported their visions, which were noted by Janice Edkins at the volunteer search center in Fort McMurray: readers of tea leaves, cards, and map energy that showed heat centered over the area from Anzac to the Christina Basin. A fisherman on Winifred Lake reported seeing a single-engine Cessna circle half of the lake around 11:00 AM on May 19, and then continue to the west. Janice knew if it was her brother, the maneuver would make sense. Winifred is the largest lake in the area and he could have used it to get his bearings. "Please keep looking," Janice urged people. "Don't give up. The search hasn't ended." The Edkins family again and again expressed appreciation for the outpouring of support. "It's been astounding."

Marion could stand it no longer. She found Stan Edkins's phone number and relayed on his answering machine the messages she'd received from his son. Within two hours Stan returned her call. "Give me the exact location. I'll send a couple of planes out to search." But nothing was found.

On June 22, Marion and her husband borrowed a quad and truck and drove out to Philomena. They unloaded and again traveled along the CNR track to Mile 153. A grizzly bear met them. "We must make ourselves known to the bear," Alex said, and began to holler while Marion blew her whistle. The bear stepped aside, then came back on the tracks after they'd passed, looking after them.

The strong feeling returned when they reached Mile 154, so they decided to follow every cutline and trail possible branching to the right of the railway track. Each ended with impassable water—a creek, a river, muskeg. They continued to Mile 158. Nothing. They turned back to Philomena, then drove northeast following the trails and cutlines in that direction. Nothing. Go back home. They'd tried.

The end of June was coming around, and still Stan Edkins and his crew of volunteers retained hope. "We're not giving up by any stretch of the imagination," Stan said bravely to reporters from the *Slave River Journal*. "Gavin has the advantage of reasonable weather as long as he hasn't been too severely injured." He recalled tales of other northern pilots who'd survived bush crashes for months, even during the winter. "If we have even a one-percent chance, all we have to do is nail that one percent and we have him back." [4]

Bruce Restou, a helicopter pilot, agreed to fly Marion and Alex over the area near Philomena in a flight authorized by Stan Edkins. Leaving at 1:00 PM on June 24, with Alex on the left side of the helicopter and Marion in the rear seat at the right, they flew to Mile 159, east and west, then worked their way south. When they spotted three lakes, a large round one and two smaller ones, the last shaped like a canoe, the helicopter turned to the left. "What is that?" Marion cried. "A white patch in the lake!"

"Just foam," her husband replied, leaning forward to look. "Nothing but foam."

But as Marion lay down at home that evening, she again saw the white patch. It was an object in the water, not foam on the water, she was sure of it. She got up and phoned the helicopter pilot. "I'm sure that wasn't foam I saw in the water," she said. "Please go back and have another look." He promised he would.

The dreams, the dreams. "You asked me if I was dead or alive," Gavin said. "Now you know. You did your share—now it's up to the others to get me out."

Marion awoke. Something had ended. It was like a funeral where one feels sadness, but also closure.

A feeling of disappointment overwhelmed the civil searchers. John Payne, who'd initially been hired by Stan Edkins to fly the Norseman and then had volunteered as a spotter for the CARES/CASARA Edmonton group, felt compelled to comment on the "lack of coordination and direction." He said, "We knew we were searching for a Cessna 150, red on white with one occupant, and that was it." The civilian crews reiterated that they'd felt incidental to the operation rather than a part of it.[5] The SAR's Op Report begged to differ. It stated that seventy-six sighting reports were received and investigated by search headquarters, and that each was retained on file at RCC Trenton.

The helicopter pilot reported back to Marion on June 27. He'd flown back over the area and hadn't seen anything. Now what? Perhaps he hadn't gone over when the sun was reflected from just the right angle, or at the right height, for the object to show. She was sure now that it was a left wing sticking out of the water. She had to tell Stan, "If you fly northwest at about 2:30 P.M. at the right altitude, you'll see the white image that I saw in the lake. It looks like a wing. Phone Bruce. He knows the place."

The dreams returned the next day. The Cessna was struck by lightning, going down in flames.

"You're gone, aren't you, Gavin?"

No reply.

Bruce Restou reported back to Marion that he'd sent Stan a map of the lake, and that he was going to get a helicopter and divers.

Stan was following up every possible lead. By now they were searching outside the flight path, in areas east and south of Lac La Biche. The next plan was to search the general area west of Lesser Slave Lake. Gavin's course could have varied to such a degree that he might even have crossed the border into Saskatchewan.

Ron McCaw, the Wetaskiwin airport director who'd also served as director of the Northwest Territories Civil Air Search and Rescue team that had successfully commanded two searches in the North, stated, "As a search commander my first thought would be to draw a radius around the storm, calculate his fuel and try to pin down his location." After conducting a grid search, the team would backtrack to check out gullies and rivers.[6] Bush pilot

Jim McAvoy reasoned, "He could have gone through ice. At that time, May, there would still be ice on lakes but it wouldn't be too good. When you land out of control the aircraft picks up speed, tightens up, and when you hit, the aircraft comes apart. The engine could be several feet down into the ground."

The Fort Smith community held a benefit concert, which raised money to pay for more hours of aircraft fuel, rentals, and mechanical inspections, and other expenses, such as food and lodging. "Stan says it's not the money that's important," the *Slave River Journal* reported on June 25, "and he won't let it stand in the way of finding his son. 'A life is worth a hell of a lot more than any money,' he says." The chances of both running out were becoming ever more evident.

Marion's Ukrainian ancestors had taught her to paint Easter eggs, pysankys, *with intricate designs symbolic of life resurrected. She decided to paint some now and started with her black paint. But the black faded on one side, and the images of the three lakes became superimposed. In the center of the third and smallest lake was a white patch. A wing. Forty days had passed since Gavin's disappearance. Perhaps his soul had been released.*

When the images of the three lakes returned in Marion's dreams, she gave them names. The largest would be called Grandma's Lake. The second, about a quarter-mile distant, would be Joe's Lake, as she'd learned through her dream that a young man and an old trapper had drowned in this lake, and their bodies had been there for many years. Let them rest. The third lake, surrounded by marsh reeds, would be Swampy— the one that contained the wing.

By July 19, Marion had had enough. She continued to see images of Gavin, walking tall, with pride. He wore a wedding band on his pinkie finger. He had two necklaces, one with an airplane pendant and the other with an eagle. She saw he was wearing the airplane pendant the day he'd disappeared. She later phoned Gavin's wife Sonia to ask and was told that was true.

Marion also visited a hypnotist. He took her back to June 24, the day she'd been flying in the helicopter and spotted the white object in the lake. It was a left wing.

On July 20, Stan Edkins called the Kuziemskys. The conversation became one of confusing non sequiturs, and ended with Marion telling Stan she'd had an appointment with a hypnotist and Stan calling her a dingbat. It was now up to Stan, Marion thought, to decide whether he wished to remove the body from Swampy Lake or leave it there. In any event, the Edkins family must exit from her waking and sleeping hours. Their voices haunted her, Stan's habit of saying "Good gracious!" and Gavin asking her to rescue him from the depths.

"Go away!" she cried. "All of you!"

Marion went to church, lighting candles for each of the Edkins family members. Gavin's candle gave off a bright tall flame from its white holder, for purity and innocence, Marion thought. Then she lit a candle for the Edkins family, placing it in a blue holder, blue for health and relaxation. The flame flickered and died. She removed the candle and placed it in a green holder, for hope, strength, and healing. The flame burned clean and bright.

For the next two nights, the dreams stayed away. Then, on the night of July 23, they returned, fierce and wild. Stan was after her, accompanied by a woman with light brown hair. "You didn't tell me everything!" Stan accused.

"Come back when I'm awake!" Marion cried in her sleep. "You are out of my dreams! You can search for Gavin until you've gone through two million dollars—but if you go to the canoe-shaped lake you'll find him. Or in the round lake just to its west." Stan nodded, and the dream faded as dawn lightened the window.

A number of weekly newspapers printed an appeal by Sonia Edkins, accompanied by Gavin's photograph: "Dear Editor. I hope that you can help me. My name is Sonia Edkins of Fort Smith, Northwest Territories." After describing the incident, Sonia asked people who worked in industries such as fishing and farming, and those partaking in recreational fishing, camping, or hunting to keep watching for Gavin and his plane. "We, my two children—a four-year-old and a twelve-month-old, both girls—are hoping that you may be able to help us at your end. My four-year-old keeps asking,

'Will my Daddy ever come home?' If you think you may have seen a small plane in trouble on May 19th, please call. Thank you. God Bless. Sonia Edkins."

Gavin's voice, from far away, begged Marion to complete her mission. "My family doesn't believe I'm dead!" The next day she phoned Stan and left a message. There was no return call.

In August, the Yellowknife aviation community, through the NWT Sport Parachute Association, organized a "Find Gavin Edkins Raffle," hoping to sell fifteen hundred tickets and raise over ten thousand dollars. The Canadian Imperial Bank of Commerce also set up a fund and received $2,600 in donations. "Our main concern is that the whole world hasn't written Gavin off," stated Paul Curren of Yellowknife. "Theoretically, he could last for months, even under winter conditions. So there's a real potential he's alive."

It was no secret that Stan and his family were under terrific strain, emotionally and financially. It was now ninety-seven days since Gavin's Cessna had disappeared, and Sonia Edkins acknowledged that "it would take a miracle for him to be found alive." A scrapbook chronicling the search was compiled for the children to read when they were older. "It will help them learn how hard people tried to find their father, no matter what happens," she said, adding that their four-year-old daughter, Shadow Lynn, still cried for her daddy.

New Theories and Evidence

In September 1996, four months after the airplane had been reported missing, a fisherman on Grist Lake (south of Winifred Lake in northeastern Alberta and between Lac La Biche and Conklin) informed the Edkins family that an image had appeared on the screen of his fish finder that resembled the fuselage of a small airplane. Grist Lake was fifty miles east of Gavin's flight plan, and certainly an area of interest. There was a summer grass airstrip at the north end of the little lake, and a winter ice airstrip on the lake close beside the found object. The RCMP brought in a team of six divers who covered an area "the size of two football fields," using sonar and a diving plane board. While the sonar identified a fuselage-shaped object four times, it turned out to be a dense school of baitfish, they said, and white shale rock. No other evidence was found after a two-day search.

On September 8, a fisherman on Winifred Lake reeled in a hank of hair which he saved and turned in to police the next day. The Edkins family found out about the discovery in December, through a chance remark from another pilot. "No one contacted the family!" Stan Edkins said to *Slave River Journal* reporter James Caroll on December 10, 1996, adding that he'd spent weeks "throwing money away" searching between Red Deer and Lac La Biche after police had found the hair in the northern lake.

Sergeant Scott Stauffer of the Fort McMurray RCMP detachment reported that preliminary tests had been done in October on the hair samples, which had determined that they were from a human being. Upon notification, the Edkins family supplied samples of Gavin's hair, which seemed to match the color and length of the samples found in the lake, but Sgt. Stauffer said it would not be possible to do conclusive DNA tests as "live root cells" were required.

The discovery of the hair in the lake opened up another mystery—if it wasn't Gavin's, whose was it? "It's an issue we're going to have to deal with," Sgt. Stauffer said, acknowledging the possibility of *another* body resting at the bottom of the lake. "But if we don't have a missing person . . . it puts us between a rock and a hard place," he said. "It could be someone's toupee that fell into the lake years ago."

Because of the lateness of the season it wasn't possible to send divers in until the spring. Stan was livid. "This is the first real proof! It's from the area where Gavin would have been flying, and the police don't seem to be taking it seriously," he stated to reporters. By the end of that fateful year, with the Edkins family's personal funds run dry and the volunteer crew dwindling, hope that the wreckage might be spotted by chance was all that remained.[7]

Lessons Learned

On April 18, 1997, Stan Edkins received a letter from Lieutenant-Colonel Keith Gathercole, Deputy Commander, Air Transport Group Headquarters, Canadian Forces Base Trenton. He expressed sorrow about the unsuccessful search for Gavin, and also for the criticism voiced by various local people regarding the Department of National Defence's (DND) Search and Rescue program.

> Every year we handle something like 8000 cases and fly something like 1200 missions ranging from simple investigations of ELT signals to full-blown searches such as that for Gavin. The expectation is that every mission will be treated with equal seriousness and that every mission will be executed to the limits of the aircraft's and crew's ability. With few exceptions we are successful in locating the object of the search and if there are any survivors we will do everything humanly possible to rescue them.

He noted that the comments received would be taken seriously and incorporated into future search master courses. They included being "open and accessible to the next of kin of the people we are endeavoring to locate and rescue," and sharing information. Flexibility in developing a search plan would be the order of the future. While one person—and one person only—must be in charge to minimize conflicts, Gathercole stated, "Ultimately, we need to listen to each other and to work together as a team in order to get the job done and do it properly."

By this time, the case was being examined from all angles. A Halifax (CP) article published in the *Edmonton Sun* (May 5, 1997) titled "Improved rescue locators eyed for aircraft" quoted a controversial defense department study that had found that "standard emergency beacons for aircraft fail to function properly in up to 75% of all crashes. When beacons do trigger, it's a false alarm in four of every five cases—costing about $2 million each year in wasted rescue flights." The Gavin Edkins disappearance was a case in point. "Sonia Edkins knows the cost of a failed emergency beacon. Her pilot husband, Gavin, disappeared May 19, 1996, in his single-engine plane in northern Alberta. The 36-year-old pilot's emergency transmitter didn't go off … Aviators and airlines cling to the outmoded emergency transmitter system because it's cheap—between $200 and $300 for a unit—compared to 10 times that amount for a more sophisticated version." Everyone, it seemed, had learned something from this tragic experience.

On May 19, 1997, a memorial service was held at St. Joseph Cathedral in Fort Smith for Gavin Edkins, with donations forwarded to a trust fund for Gavin and Sonia's children, Shadow and Angela. On June 5, 1997, Stan Edkins had the sad duty of signing an affidavit to be presented in the Supreme Court of the Northwest Territories in accordance with the Presumption of Death Act. The paper stated that Stan had participated in the search for Gavin, along with the Department of National Defence, from May 19 to June 2, 1996, and continued six weeks thereafter with local search crews. No evidence of Gavin or the airplane was found, and no one had heard from the pilot from that day, May 19, to this. It was the most difficult document he'd ever put his name to.

Alex Kuziemsky died on May 27, 2002, and now Marion dreams of him. And of Gavin too, still under water. Her family has joined her quest to find the lost pilot and they've ventured out to the Philomena area about twenty times. Her son-in-law bought a metal detector that can trace metals down sixty-five feet and with that they'll continue to check Grandma's, Joe's, and Swampy lakes. Whenever she goes there, the smell of shaving lotion permeates the air.

Marion's son took her out on the lake in a canoe in 2002, and the grizzly bear once again joined them, coming to within fifty steps of their canoe. "Should I kill him?" her son asked, reaching for his gun. "No, he's not meant to be killed. He's here to help us," Marion replied. The grizzly silently watched them for a few minutes, then ambled off.

By that time Marion was sixty-one years of age, a grandmother, and not in the best of health. But she could not give up. Her dreams wouldn't allow it.

Stan Edkins cannot give up either, even though the search cost him his business and a radical life change. He concluded his aviation charter company and in 1998 sold his 1938 Norseman to the Canadian Bushplane Heritage Centre in Sault Ste. Marie, Ontario. Stan and his partner left the North and now reside west of Edmonton.

The Edkins family published a thank-you letter in the Fort Smith newspaper in which they hoped to include the names of all individuals, companies, and organizations that had helped, but the task was nearly impossible. The two-page list included over two hundred Canadian volunteers and a number from the United States and overseas: "Your actions, through giving of yourselves, physically, emotionally, materially and thoughtfully, is touching our lives forever . . . The following list is not complete—and the search is not over."

THE GLITTER OF GOLD

The McAvoy family were never strangers to the north country. The boys—Jim, born September 17, 1930, and Chuck, born October 5, 1932—and their sisters Eva and Goldie (the latter so named as the first white child born in Goldfields) grew up in places such as Fort Chipewyan and Goldfields. When their father, also named Jim, became involved in exploration and mining in Yellowknife, NWT, Jim and Chuck followed. There they worked in diamond drilling, seismic, and prospecting, and became familiar with the gold fields of the North.

Air service was scant in Yellowknife in 1944, and so the McAvoy Diamond Drilling & Development Co. Ltd. bought three successive Fox

Moths for use in the business. When the pilots they hired wrecked them all, it was time for the family to take charge. Young Jim left for Edmonton in the winter of 1948 to take flight training with Western Air Motive for private and commercial licenses, and endorsements on multi-engine, floats, night flying, and instruments. The McAvoy company's airplanes stayed around longer with Jim at the controls. Chuck soon followed his older brother and took flying lessons as well.

While awaiting an operating certificate for their own charter air service, Jim and Chuck would often fly to the Nahanni area, 350 miles southwest of Yellowknife, in their Cessna 180 outfitted with wheel-skis. They'd drop down to visit with prospectors and trappers and, for fun, watch for any old plane wrecks to check for cargo that might be left behind. Perhaps they'd come upon some American cash once destined for army payrolls, or gold.

The Nahanni, also known as Deadmen's or Headless Valley, can be a dangerous place for the unwary, or the unlucky. "People are pretty careful when they get into that area," reflects one pilot who became familiar with the Nahanni's sudden weather changes and unforgiving mountainous territory, and lived to tell about it. "Some go in and never come out." Numerous aircraft parts strewn on mountains or at the bases of rocky cliffs, along with scattered bones and baggage, affirm such words of caution.

The McAvoys, blessed with colorful personalities and a solid background in gold prospecting and mining, came to know most northern residents. On May 7, 1960, the brothers flew to the Nahanni carrying some mail for prospectors and trappers, including three men who had set up camp on McMillan Lake in the Flat River area.

The McAvoys had heard Dean Rossworn, Orville Webb, and Thomas Pappas announce the previous year that they were leaving Yellowknife to go west to search for the legendary gold of the Nahanni. They were heading into territory unknown to them and staying over the winter with the idea of using explosives to "go down below the frost" to open up a gold mine.[1]

They'd scoffed at the curse of the Nahanni, where in the early 1900s, head-less bodies had been found of gold prospectors Willie and Frank McLeod and their partner Robert Weir. Partial skeletons of other prospectors who'd come to pry riches from the valley's gold-filled streams had provided the basis for a variety of macabre names, such as Deadmen's Valley, Headless Valley, the Valley of Vanishing Men, and for its principal river, the South Nahanni, the River of Mystery. But that was old stuff, the Yellowknife prospectors thought, exaggerated over the years.

The previous September they'd chartered Max Ward's new single Otter to fly them, their dogs, equipment, and supplies—sufficient to last until mid-March—from Yellowknife to their mine site before freeze-up. A week later, on October 1, a chartered Wardair Beaver aircraft brought two more men, John Richardson and Alex Mieskonen, with tools and equipment to do sim-ilar work. The first group of three occupied a cabin with a lean-to at the back, while the second group of two moved into a smaller cabin, with both sharing a high cache to store supplies.

They knew there was game in the mountains that would stretch out their supplies and feed their five dogs. When an early snowfall drove a small herd of caribou down to the newly frozen lake, they shot three animals and pre-pared the meat, believing that was allowed under their prospector's licenses. Then they began the work for which they said they'd come: to blast and enlarge the mine entrance and follow the seam by drifting and tunneling.

On January 3 they were visited by Frank Bailey, the game management officer from Fort Simpson, who said their prospecting licenses did not allow them to hunt big game except in cases of starvation. He confiscated half the meat and left with a warning that he might return to take the rest unless he found the men starving. With that, he took off in the chartered Beaver air-craft that had brought him there. Bailey returned a few days later on January 7 with RCMP Constable Victor Werbicki. After checking whether the men had sufficient supplies to last until their pickup aircraft arrived in early March, he confiscated the rest of the meat.

Chuck McAvoy (right), discussing business with Gordon Bartsch, the pilot he'd hired to fly his newly refurbished DC-3 (CF-JWP). (Courtesy Gordon Bartch)

March had come and gone with bad flying weather. Perhaps that's why no plane arrived, the prospectors reasoned. However, when the weather grew warmer and the days longer, there was still no airplane. They'd shot a few muskrats, but there were no more caribou. Soon they were existing on squirrels, birds, rotten potatoes, and their dogs. When the ice began to break up, an airplane would not be able to land with supplies or move them out. They tramped out an SOS on the lake before the ice went out, and then sat down to discuss the increasingly serious situation.

Rossworn, Richardson, and Pappas set out on a fishing trip to Clark Lake, which was unsuccessful. When Rossworn returned to the cabins, he found only Mieskonen there, who said Webb had gone out hunting. "Alex Mieskonen had been acting strangely for some time," Rossworn reported. "He told me he was going to take a walk up the snare line. After he left, I lay down as I was tired from my trip to Clark Lake. A short while later I heard an explosion and thought Webb must have found some game." Suddenly Webb burst into the cabin, his face pale, his eyes wild. "Alex has blown himself up with dynamite! Out there on the north end of McMillan Lake!" The blood-spattered snow told a grisly story. Alex Mieskonen had strapped dynamite to his belly, attached a cap and fuse, and lit it, blowing himself to bits. The two partners wrapped the remains in a tarp, and tied their remaining dogs nearby to guard it from wild animals. On April 7, when the snow had melted and the ground became less hard, they buried Alex Mieskonen.

By April 12 the situation was desperate. During the previous month, all the prospectors had eaten was soup from a boiled caribou hide and some old marrow bones, alder and birch roots, squirrels and rabbits when they were lucky, and the rest of the dogs. They decided that someone had to leave camp and seek help. Rossworn and Richardson would stay behind to wait for a plane, while Webb and Pappas would hike to Nahanni Butte, at the junction of the Nahanni and Liard Rivers, one hundred miles distant. The trip would involve snowshoeing through eight- to ten-foot depths of soft snow and crossing rivers where the ice was breaking up in country unfamiliar to them. But from there they could send word to Fort Liard or Fort Simpson. The two men took a can of cooked dog meat, some tea, salt, butter, cooking fat, a tarp and blanket, two .22 rifles and a .30–.30 rifle and ammunition, and said their farewells.

No airplane arrived at McMillan Lake, only increasing hunger and worry. They had killed two dogs that had become sick, and three others had already been slaughtered for food. One sick dog, whose carcass had been thrown into the nearby creek, was now retrieved and consumed by the two starving men. Rossworn and Richardson were at death's door when they heard the roar of an engine.

On May 7, Jim and Chuck McAvoy looked down from their aircraft to see SOS tramped in the snow on McMillan Lake. They circled and dropped down to land on the rapidly thinning lake ice. Two men stood outside the cabin door and it was obvious that something was terribly wrong. The men looked bushed, a little crazy, and gaunt as if they hadn't eaten in weeks. As the men wolfed down the sandwiches the McAvoys had on board, they told the pilots their story. When Jim and Chuck went outside to check out the dismal camp, they noticed a large wooden box, two feet by three feet and open at the top, in the lean-to behind the cabin. They peered inside and grinned knowingly at each other. Being miners, Jim and Chuck recognized the contents: gold ore concentrate from the hard-rock mines of the Northwest Territories.

They'd heard the prospectors had a devious plan, and here was evidence. "They'd been employed at the Discovery Mine, 52 miles out of Yellowknife, and had skimmed it off the mill as the gold went through to be separated from the hard rock," Jim recalls. "Everybody did it. Everybody high-graded. When miners left on the airplane to go 'outside,' they often carried 'empty bags' with them. But these men hadn't wanted anyone else to carry their bags, which were already full. When they returned, the bags would be filled, of course, with bottles of whiskey and other supplies."

The prospectors' plan was to cache the concentrate, then spread the word that they were coming to the Nahanni to search for the lost McLeod gold mine. They'd bring the concentrate with them, salt the frozen creek beds, then, *voilà*! There it lay in the spring meltwater, just waiting for the lucky prospectors. "It was a crazy idea," Jim McAvoy says. "Anyone who knows gold knows placer and hard-rock gold are completely different—one is sharp-edged, natural, and the other has been ground out through the milling process. But, that was their idea, and here was the box of high-graded, gold-filled concentrate behind the shack."[2] The McAvoys said nothing, just went back to the cabin and rejoined the two survivors who were still busy devouring the sandwiches.

Intensive searches were made by land, river, and air for the two partners who had snowshoed out. Their bodies were never found. During the court hearing held later that year in Yellowknife, the story took some twists and turns regarding pickup arrangements. According to one version, Chuck McAvoy had been chartered, which he vehemently denied in his statement to the RCMP. "I did not know about their arrangements for going in. I told them I would be going into Nahanni before spring and would call in to see them [but] I had no charter." Chuck added that, in his opinion, the men were nuts to prospect in these mountains in the winter, and hinted that there must have been another reason they were there. It was proved that Chuck McAvoy had not been chartered, only that he'd said he would fly in at some point to deliver mail. It was then revealed that another company had been chartered—it was found written in their books. Why they hadn't come became the question.

The game warden was also found not at fault. He'd left some meat, confiscated the rest, sent word for a plane, and then, satisfied they had enough food to last, took the rest as he felt justified in doing. "Frank Bailey was known as a man who followed the letter of the law," Jim McAvoy says. "He had to do what he thought was right." The prospectors left too much to chance and the airplane just didn't come. There was no assurance that there was sufficient game to sustain them and their five dogs through the long hard winter.

Jim McAvoy adds an addendum to the story: when he returned to the men's campsite some time later, the box of concentrate he and Chuck had seen behind the cabin was gone. Its presence, and absence, was never reported. Men come and go, but the McLeod mine in Deadmen's Valley is still holding onto its gold, and its legend.

The McAvoy brothers' charter licence came through and McAvoy Air Service came into being. They were extremely busy, but conflict grew between the brothers over bookkeeping records, the level of aircraft maintenance, and Chuck's risky flying. When they split up the business, Jim estimated that Chuck would last a year based on his methods of operation. "The writing was on the wall. He had been lucky up to then. He'd fly

Jim McAvoy, with his poodle Salty looking out the window of deHavilland Single Otter (CF–CZP). (Courtesy Jim McAvoy)

aircraft when other pilots wouldn't go."

In June 1964, Chuck and two American geologists, Douglas Torp from Duluth, Minnesota, and Albert Kunes from Prentice, Wisconsin, were reported missing in a Fairchild 82 (CF–MAK) on a flight to the Roberts Mining Company's gold claims, 230 miles north of Yellowknife. When they left from Bristol Lake at 2:00 AM on June 9, 1964, the weather was clear but soon deteriorated. At last report, Chuck indicated he was heading for Itchen Lake, east of Great Bear Lake. The aircraft was not seen again. A massive search spent six hundred hours checking over two hundred thousand square miles, to no avail.

Thirty-nine years later, on August 4, 2003, Jeff Constable, a pilot with Great Slave Helicopters flying a team of geologists from Ashton Mining, spotted the burned-out wreck of the Fairchild. It lay on the rocky terrain just south of Bathurst Inlet on the Northwest Territories/Nunavut border,

Thirty-nine years later, the remains of Fairchild 82 CF-MAK at the crash site. (Courtesy Ray Kaduck)

340 miles northeast of Yellowknife. The aircraft had landed short of a lake, hit hard on the boulder-strewn surface, and caught fire. The post-crash fire from exploding fuel tanks in the wings and cockpit area completed the tragedy. The remains of pilot Chuck McAvoy (as well as his perfectly pre-served wallet and its contents) and those of his passengers were found amid the wreckage. Memorial services were held in both Yellowknife and Edmonton. Later, all of the remains were gathered and cremated en masse, with an on-site memorial service planned for sometime in the future.

Joe McBryan, owner of Buffalo Airways (Yellowknife and Hay River, NWT), and a lifelong friend of Chuck McAvoy, has removed CF-MAK and plans to have it restored and displayed. He reflects the feelings of many in the northern aviation community when he says that finding McAvoy and the Fairchild was, as written in the *Edmonton Journal* on August 15, 2003, "more important than if they'd found the Franklin Expedition." One of the lost was now home.

THE HAUNTED FLIGHT PATH

Stretching east-west between Vancouver and Penticton, BC, is an area notorious for receiving wrecked planes, long known as "the haunted flight path." The peaks of the Coast and Cascade Mountains vary from five thousand to nine thousand feet and dip deeply into fog-filled valleys. Powerful Pacific winds and unpredictable weather patterns can bring whiteout snowstorms that can give way to bright sunshine within minutes.

This five-thousand-square-mile wilderness area, once known as the most treacherous flying country in the world, lies along the western edge of a ten-mile-wide flight path formerly known as Green One, which is Canada's principal airway from Pacific to Atlantic. Three mountain peaks within a fifteen-mile radius south of Hope, BC, where a total of eighty-six people died in plane wrecks, have been sealed off by the BC government. Here, thirteen people met their deaths in a Canadian Pacific Air Lines (CPAL) Lodestar in 1942; eleven in a RCAF Liberator in 1945; and sixty-two in a Trans Canada Airlines (TCA) North Star in 1956. Each crash site has been designated as a wilderness cemetery and is marked by cairns.

THE LODESTAR

The first in this cluster of calamities involved a CPAL Lockheed Lodestar. The twelve-passenger civil airliner (originally registered as CF–TCS) had recently been purchased from TCA, and since November 17, 1942, it had carried the registration of CF–CPD.

On Sunday, December 20, 1942, the aircraft was en route to Vancouver from Whitehorse, Yukon, via Prince George, BC, carrying ten passengers and a crew of three. At 5:30 PM it simply disappeared. "Mystery shrouds its fate," stated the *Vancouver Daily Province*, as the air traffic control center at Vancouver International Airport tried to comprehend what might have happened in the last fifteen minutes before the aircraft was due to land, and where it might have gone. When no further reports were received from CPD, the worst was feared.

The Lodestar's pilot, Captain Ernest Kubichek, was a veteran northern flyer with at least ten years of mountain flying experience. He had taken his flight training in Calgary and then flown extensively throughout northern

A Lockheed 14-H2 Super Electra (CF–TCS). The Super Electra replaced the Lockheed Lodestar at Yukon Southern Air Transport in 1942. (CPAL collection)

Alberta, British Columbia, and Yukon. First Officer W. J. (Bill) Holland had recently joined CPAL following two years with the Trans-Atlantic Ferry Service. This was his familiarization flight to the Yukon Territory. The stewardess was Edna Young from Edmonton. The flight carried Canadian and American residents coming from Alaska, the Yukon Territory, and northern British Columbia.

Captain Kubichek and First Officer Holland had filed a flight plan to follow the north beam into Vancouver, the usual route from Prince George. This would bring them over Gun Lake in the Bridge River district, past Garibaldi and Squamish, and over the Howe Sound mountains into the city. Flying at twelve thousand feet on the beam, they last reported over Squamish. But no plane came in.

Under the coordination of CPAL's general superintendent, Captain Herbert Hollick-Kenyon, searches were mounted by the airline's search and rescue unit, military and civilian planes, and troops from the Royal Canadian Engineers. As poor weather hampered their efforts, Hollick-Kenyon interviewed members of the public to gain any information on the missing aircraft.

More than one hundred people reported seeing or hearing an aircraft circle for forty-five minutes over Chilliwack and in the upper Vedder River valley before it disappeared into the night. Some said there were no lights on the aircraft. Could this mean there had been a power failure, thus their inability to radio in? With three radios and an emergency power unit, it seemed unlikely. Perhaps the pilot was seeking a landing strip or attempting to orient his position. Had CPD, due to poor weather, turned east to find a smaller but more accessible landing strip up the Fraser Valley? Had the pilot, while seeking his bearings in a cloud-filled sky, picked up the radio directional beam from Princeton and flown east on that beam across the mountains? Several days later, a man living in the Surrey area near Vancouver reported that he'd been listening on his radio short-wave band at the time of the Lodestar's disappearance when he'd heard the words, "I can't come in!" followed a few seconds later by, "I am going to crash!"

Where was the airship? Captain Hollick-Kenyon had seen almost every situation in his long aviation career, but this was absolutely baffling. He believed these people had indeed heard the Lodestar, as it was the only aircraft in the air in that district at that time. But why was it in the area of Chilliwack after it had last reported from Squamish? CPAL offered a five-hundred-dollar reward for information leading to the discovery of the missing aircraft and mounted "the greatest manhunt this district has ever seen."[1]

The RCAF began a detailed search from the US-Canadian border to Garibaldi, with eight reconnaissance planes flying three miles apart. Later, searches were expanded south of the border. Soldiers from the Royal Canadian Engineers training center at Vedder Crossing "pushed through the rough undergrowth and crawled over fallen trees along the ridge north of the upper Vedder River."[2] Civilians combed rocky canyons and steep mountain slopes, while RCAF bombers and CPAL airplanes flew overhead, carrying veteran spotters and residents of the area.

The searchers who gathered for daily assignments were given warnings about the type of country over which they would be flying. "Watch those valleys in here, they're treacherous. The air may be turbulent. You may come suddenly on a dead-end and have to do a stall turn."[3] Thick, low-hanging clouds obscured vision, and pouring rain made searches extremely dangerous. The mountains near Hope had received a new blanket of snow. But out went the search planes, to be tossed like ping-pong balls by driving winds. Even the president of CPAL, Grant McConachie, was out in a twin-engined aircraft, searching for one of his own.

A reporter, Stanley Burke Jr., eager to give his story on-the-spot immediacy, risked his neck to accompany one of the searches.

> On either side at frequent intervals are long menacing sawtooth ridges and sheer-faced cliffs, from which waterfalls drop for

CPAL's Lodestar (CF-CPD) crashed on December 20, 1942, on Mount William Knight Peak. (Courtesy George Maude)

hundreds of feet. The upper slopes are blanketed in snow, the lower ones wet with slush and rain. Later in the year, the survivors might be able to make their way down the frozen streams, but these are not yet frozen over and they would be forced to make painful progress through the heavy underbrush.[4]

Christmas was celebrated with further searches, and hope from families, friends, and aviation personnel that the search would end quickly with all passengers and crew found alive. But it was not to be. The aircraft was not found until eight months later, accidentally sighted by CPAL Captain Don Patry while flying another company airliner. The Lodestar had inexplicably veered to the southeast on its way into Vancouver and struck the peak of Mount William Knight in the Cheam range. The thirteen men and women aboard the aircraft had all lost their lives, and no bodies were retrieved. The reasons for the sudden reversal of the flight plan and the resulting crash have never been learned.

THE LIBERATOR

The second major fatal accident to occur in a fifteen-mile radius mountainous area south of Hope involved a RCAF Consolidated B-24 Liberator GR-VII that went missing two and one-half years later, on June 1, 1945. Western Air Command stated that the four-engined bomber, with eleven Royal Air Force personnel aboard, had left Abbotsford at 9:00 AM for a routine cross-country training flight over the interior of BC. A radio call sent to the aircraft at 9:30 AM had received no reply. A transmission from the Liberator, due when it should have passed over Penticton at 9:45, was not made.

Even so, it was expected that the Liberator would return to its Abbotsford base on the Lower Mainland shortly before noon. It didn't come in. No sighting reports were received from anyone near the Coquihalla Pass on the east-west radio beam, which was monitored at evenly spaced intervals by railway section hands. If the aircraft had gone off course, it could be anywhere in the heavily forested, craggy, and virtually inaccessible mountains.

At noon on Friday, June 1, ninety minutes after the bomber was reported missing, radio signals began to be heard at half-hour intervals from a region known as the Chilliwack Hump, sixty miles by air southeast of Vancouver and among mountains called Cheam, Silvertip, and Silver Peak. The signals, thought to be coming from a portable set carried in the Liberator, brought the focus to a ten- to twenty-mile-wide section southeast of Mount Cheam, bounded by Elk Mountain, Lady Peak, Chilliwack River, and Mount Tomiki. The signals continued until Tuesday, June 5, when they ceased at 3:30 PM.

This "graveyard of lost planes," named because of the proximity to the wreckage site of the CPAL Lodestar and several other aircraft, taunted searchers with impassable weather. But when weather conditions permitted, RCAF members made daytime searches in light aircraft, in addition to night patrols that looked for flares through breaks in the clouds. Units from the Pacific Coast Militia Rangers and BC Forest Rangers were led overland by veteran game wardens, guides, and Alpine Club members. Parachutists, woodsmen, and other survival experts stood by at various bases, ready to help with search and recovery.

"Signals Hint Bomber Crew Alive in BC 'Air Graveyard'," read headlines in the *Vancouver Daily Province* on Wednesday, June 6, 1945, painfully recalling for readers and searchers the recent discovery of the Lodestar on nearby Mount William Knight. Indeed, short-wave radio signals were picked up at Chilliwack by Pacific Coast Militia Rangers on the international distress call frequency at five hundred kilocycles, augmenting signals received earlier by Western Air Command. The RCAF set up a radio and direction-finding station at the four-thousand-foot level of Elk Mountain, but somehow doubted the ghostly radio signals came from the missing bomber. Could the men still be alive, seven days later? It was possible. On board were full rations for five days, as well as sleeping bags, a Gibson Girl radio set, medical supplies, fishing gear, and a Very pistol and shells.

"Seven injured . . . still alive," came three weak messages on June 8! The receiver, J. D. Usselman, a Chilliwack resident, reported the message had come by telegraph signals and radio voice, in the weak tones of a man speaking with a British accent. Usselman had produced the words by combining all the messages received. Although the RCAF was not convinced of the authenticity of the messages and the technical ability of the homemade receiver to pick up messages on the five-hundred-kilocycle frequency, anything was better than nothing, which was all they had to go on before. They quickly broadcast instructions to any airmen who might be alive amidst the wreckage, and the search refocused on a canyon between Tamihi and Slesse Creeks.

At first light on Saturday, June 9, fifty searchers and six additional Pacific Coast Militia Ranger parties converged on Tamihi Creek, a tributary of the Chilliwack River, and about sixteen miles from the city of Chilliwack. Ground haze limited visibility, and rain and heavy overcast skies obscured the mountaintops above the five-thousand-foot level.

"Sudden death lurks in this wilderness for the airman who defies the elements," wrote *Vancouver Daily Province* reporter James Fairley, who flew over the area on June 8, searching for freshly snapped tree limbs during the first break in the weather in seven days:

> We flew into a frightening mass of black storm clouds [and] I realized how a pilot must feel when the weatherman gets vicious by producing black clouds from nowhere in an aircraft's path . . . Southeast of Chilliwack, flyers play leapfrog over mountain peaks that reach higher than 7,000 feet into the blue—when there is any blue.[5]

Rumors of sightings continued to come in: three green Very flare-lights were spotted on a razorback on Cheam Mountain; a number of shots were heard from the Tamihi Creek area; smoke was seen from another mountain east of Chilliwack; tracks were spotted on a snowy trail, a parachute was found on a mountainside, and wreckage on a glacier southeast of Hope. Was this *another* Liberator that had been lost from Abbotsford air base last November 10, 1944, with a crew of ten flyers? Or was it abandoned mine buildings whose metal roofs might glint in the sunlight and appear as a wrecked airplane? Could the aircraft have become buried in the twenty-foot snowdrifts that now filled some valleys? All clues were investigated.

On June 11, a sighting of the Liberator was reported in the approximate area indicated by radio signals received by J. D. Usselman near Tamihi–Slesse Creeks, twenty-six miles southeast of Chilliwack or eight miles south of Chilliwack Lake. "There are signs of life," Western Air Command announced, based on information received from a Beechcraft pilot. "Parachute jumpers are going in." The logistics were rapidly planned for getting survivors out over mountainous, heavily wooded, wilderness terrain. Then Western Air Command's highest ranking officer, Air Vice-Marshal F.V. Heakes, AOC, came for a three-hour flight over the area. "There is no sign of the missing aircraft," he announced sorrowfully on his return. The

Western Air Command called off its ground searches on June 15.

Two days later, a RCAF rescue squad flying a Norseman spotted wreckage of an aircraft several hundred feet below the peak of Mount Foley in the Cheam range. Was it the Liberator, or one of two US airplanes that had also been reported missing in that area? They marked the coordinates and returned to find it was, indeed, the wreckage of the Liberator, just fifteen hundred feet southeast of where the CPAL Lodestar had crashed in December 1942. From the air, they could see that the Liberator had hit the mountain and exploded. The slide down the slope had stripped the fuselage of its skin, scattering wreckage over a three-thousand-foot area and leaving the skeletal framework wedged in a valley between the main peak and another smaller peak on the mountain.

During the time spent searching for the Liberator—called one of the greatest aircraft hunts in the history of Canada to that time—RCAF planes flew 100,000 miles in 372 sorties, with 114 airmen and 40 civilians and army personnel making up 14 different ground parties. Could anyone have survived? The job now was to find out. This required a difficult hike from Chilliwack River to Ford Creek, and a treacherous climb to the 7,400-foot level of Mount Foley, involving air force personnel from air-sea rescue units, Pacific Coast Militia Rangers guides, and a Royal Canadian Army Service Corps' first troop pack train, carrying everything from radio sets to snowshoes.

Heavy rain created quagmires and floods, stopping progress on the ground and turning back air patrols. A RCAF Ventura dropped further supplies by parachute to the climbers who, hidden from sight by clouds, struggled through swollen creeks and up logging trails that rose one thousand feet in five miles, deep into the home territory of cougars and bears, including grizzlies. Higher up, steep valleys, filled with snowdrifts up to forty feet deep, lay between mountain peaks. Searchers used snowshoes, then resumed hiking and rock climbing over ever-higher elevations into the thin air near the peak of Mount Foley.

The searchers reached the wreckage on June 22. All persons on board the Liberator were dead. Three remained inside the fuselage, and eight were found on the mountainside. The force of the crash had driven the tail assembly deep into the side of the mountain. The wings had ripped off and were thrown up onto the rocks. One engine had shattered and burned, while the other three had been pitched three thousand feet below the initial crash site, which had started a rock slide that carried parts of the aircraft down with it. If the aircraft had been 150 feet higher, it would have missed the mountain peak, or a short distance to either side would have brought it around the peak. Instead, it hit just above the 7,400-foot level, traveling at an estimated speed of 225 to 240 mph. The wreckage was unsalvageable, and all was left to repose on the mountain. Plans for a funeral on Mount Foley were delayed because of the weather until July 7, when a wooden cross with a RCAF ensign, bearing the names of the eleven men killed in the Liberator on June 1, was placed between Mount Welch and Mount Sill in the Cheam range. By then, the military was reeling from a fresh accident.

On July 3, 1945, at midnight, two RCAF Liberators smashed into each other on the runway at the Abbotsford air force base. This was the fourth serious military mishap to occur in British Columbia in two months, bringing the total of such fatalities to twenty-six. People wondered, with good reason and silent prayers, if the war was being fought here, rather than overseas.

THE NORTH STAR

Late on the rainy afternoon of December 9, 1956, the trilogy of major fatal accidents within a fifteen-mile radius on BC's "haunted flight path" was completed. TCA's North Star DC-4 (CF-TFD) taxied down the runway from gate five at Vancouver International Airport, ready for takeoff on Flight 810 to Calgary, Regina, Winnipeg, Toronto, and Montreal. The count of sixty-three passengers (forty-four men, fifteen women, and four children) plus three crew was complete. The ship's cargo contained twenty-seven bags of mail, one thousand pounds of freight, five hundred pounds of fresh flowers, and passengers' luggage, which included jewelry and money—lots of money.

One passenger alone, Kwan Song of the Bowery district of New York City, was returning from Hong Kong and was believed to be carrying $80,000 US in a special money belt. George Woon, a Chinese restaurateur from Cobalt, Ontario, had more to grieve than lost coin. He had paid "a substantial amount of money" to a black market organization in the Canton area to have his eighteen-year-old son, Ela Woon Yuen Way, smuggled out of Communist China to Hong Kong. From there the boy had taken a flight to Vancouver and caught Flight 810.

Although the passengers included the usual mix of families and business commuters, there were also five football players returning from a Shrine East-West All-Star game. They included Calvin Jones from the Winnipeg Blue Bombers, who had missed an earlier flight and at the last minute been given a vacant flight attendant's seat on Flight 810. On board as well were four Saskatchewan Roughriders: Gordon Sturtridge, #73, and his wife; Mario DeMarco, #55; Mel Becket, #40; and Ray Synnyk, #56. The players who had participated in the Shrine All-Star game had been insured for forty thousand dollars each by the Continental Casualty Co. of Toronto, 50 percent of which would be paid out to the club in the unanticipated event of an accident. Each player had also purchased one hundred thousand dollars in life insurance from a machine in the terminal building. Noting this, other passengers had lined up at the machines, with the result that passengers on Flight 810 carried a total of two million dollars' worth of life insurance.

The lone flight attendant on duty, Dorothy Bjornsson, had the job of explaining to boarding passengers why the flight was leaving late: a still-raging storm had delayed this aircraft on its in-flight to Vancouver. Just at the point of takeoff, a pilot from inbound TCA Flight 801 reported a towering buildup of clouds in the Cascade Mountains, with ice encountered at fifteen thousand to sixteen thousand feet. "But you should be a lot higher than this if you're eastbound," he assured.

High winds, turbulence, and ice—not a great forecast. Captain Alan J. Clarke, the thirty-five-year-old commander of the North Star, had served with RCAF Bomber Command in the United Kingdom Middle East, and was now a veteran airline pilot. He and First Officer John C. T. Boon, twenty-six, received a meteorological briefing from the weather office that augmented the westbound pilot's summary: a trowel (a trough of warm air aloft) would be encountered east of the Cascades; ahead of this, the air mass was very warm and moist, which had brought above-normal rain to Vancouver that day. A solid cloud deck existed from three thousand to twenty-one thousand feet from the Cascades to the Divide, and there were icing conditions from eight thousand to fourteen thousand feet in the mountain regions.

Captain Clarke took note of the advice and signed the flight plan at 5:14 PM. The wings of the transport aircraft were filled with 11,400 pounds of high-octane fuel, bringing takeoff weight to 76,850 pounds. Because of the weather, Clarke requested another eight hundred pounds of fuel (totaling 13,394 pounds of fuel plus oil aboard).

The aircraft departed from Vancouver's airport at 6:10 PM from runway 11. It was cleared to cross Westham Island to the south at three thousand feet, and then banked eastward through cloud and rain. At 6:28 PM it reached ten thousand feet. Gaining five hundred feet per minute, CF–TFD was soon flying thirteen thousand feet over Abbotsford and fifteen thousand feet over Cultus Lake.

At sixteen thousand feet, the aircraft encountered turbulence and dreaded icing conditions. Hearing a report from a TCA Super-Constellation pilot that cloud tops were at twenty thousand feet, Clarke climbed to nineteen thousand feet, but still was buffeted by severe turbulence. After gaining permission from air traffic control, he zoomed up to a clear level of twenty-

TCA's DC-4 (CF-TFD) taxied down the runway from Gate Five of Vancouver International Airport on December 9, 1956, to its fate on Mount Slesse. (Courtesy George Maude)

one thousand feet. Just as he neared Princeton—the eastern end of the mountain pass whose western end is at Hope, and forty-two minutes after takeoff—serious trouble began. "We have a fire," Captain Clarke radioed. He quickly shut down the number two engine, the indicated source of the trouble, feathered the propeller, and activated the fire extinguisher.

But *was* there a fire? Sometimes the North Star's instrument panel would so indicate, when in fact there was no fire. But such a warning must be acted upon, so the captain and first officer dutifully turned around the large, heavily loaded, now three-engine transport airplane. Back through the storm they flew, bucking a headwind and ever-increasing updrafts and downdrafts, to the haven of Vancouver.

No sign of fire was yet visible. Passengers sitting on the left side of the aircraft would have seen a spotlight over the number two engine (on the inner port side) that showed the propeller to be motionless and ice-coated. The sound of four Rolls-Royce Merlin engines would be reduced to three, and the turbulence was likely notable.

Five minutes later, at 6:53 PM, Clarke reported in to Vancouver's air traffic control tower. "We're endeavouring to maintain 19,000 feet. We would like clearance immediately to get down if we can. We're losing altitude quite fast here." Through static messages relayed from TCA radio at Vancouver and air traffic control, Captain Clarke was given clearance and lowered to fourteen thousand feet, following the Green One airway. He then reported that a 95-mph headwind had reduced his ground speed to about 100 mph.

"You'll be able to hold 14,000 okay, will you?" the air traffic controller asked.

"I think so," the captain replied.

"810 by Hope at 7:10," Clarke radioed to air traffic control (ATC), eighteen minutes after he'd turned back and approximately one hour after his initial takeoff. "Request descent down to 10,000 feet." He received clearance to eight thousand feet or above, and was asked to remain on ATC frequency. The absolute minimum altitude that TCA would allow its airplanes to fly along the Green One airway from Hope to Maple Ridge, twenty-three miles from Vancouver, was ten thousand feet, but because the government minimum was eight thousand, that is what he was given. The altitude choice was now up to Captain Clarke. Acknowledgment of this clearance was the last transmission received from Flight 810. Repeated calls were made to the aircraft from 7:21 PM on, but no reply came.

The radar team of the United States Air Force's (USAF's) North American Air Defense Command (NORAD), based near Bellingham, Washington, had been keeping a watchful eye on Flight 810. They were alarmed to see that the North Star was not on Green One—but twelve miles to the south where an even *higher* altitude was necessary. One minute later, the North Star disappeared from their radar screen in the vicinity of Silvertip Mountain, an 8,530-foot peak located twenty miles southeast of Hope, and contact was lost. "Your Flight 810 has gone off our scope!" the radar operators called to Vancouver air traffic control.

For the next eighteen days, fleets of military aircraft and ground parties of police and woodsmen once again participated in what would be called the most intensive air and land search ever conducted in Canada, through incredibly bad weather. Because civilian aircraft were considered inadequate to fly in such turbulent mountain conditions, the federal government forbade any but official search planes in the area. RCAF CF-100 all-weather, jet-fighter aircraft logged six hundred hours searching over five thousand square miles, with weather cutting their search time to 386 hours. Ground parties hopelessly fought their way through twenty-five-foot snowdrifts. Nothing was found.

Newspapers faithfully gave accounts of progress—or lack of it—along with synopses of other recent Canadian air disasters that were sure to put people off flying. Earlier that year, on May 15, 1956, a RCAF CF-100 had crashed into a convent of the Grey Nuns of the Cross at Orleans, Ontario, killing fifteen; on April 8, 1954, a RCAF trainer collided with a TCA North Star over Moose Jaw, Saskatchewan, killing thirty-six; on October 17, 1951, a Queen Charlotte Airlines aircraft crashed at Nanaimo, BC, killing twenty-three; on July 21, 1951, CPAL's DC-4 was lost en route from Vancouver to Anchorage, Alaska, on a Korean airlift, with thirty-eight on board, and so on.

The names of the crew and passengers on the North Star were listed, coupled with airline officials' statements. All were mystified by this disappearance. Air and ground searches were started and then delayed as winter storms continued on both sides of the Rocky Mountains.

On Sunday, May 12, 1957, Vancouver climbers Elfrida Pigou, Geoff Walker, and David Cathcart attempted to scale the multipeaked Mount Slesse. Aptly named for a Salish word for "fang," this jagged mountain rises just north of the Canada-US border in the Cascade Mountains. That day, heavy cloud caused the climbers to make a wrong turn that led them to a lower pinnacle. There they found the remains of an airplane—a big one. A piece of metal jutting out of the snow revealed the identification CA 37-3-2000-63B. When they brought their find to the RCMP, they learned it came

Aptly named for a Salish word for "fang," Mount Slesse rises just north of the Canada-US border in the Cascade Mountains. (Courtesy Evan James Bullock—DND photograph)

from the wing of the missing North Star.

CF–TFD had hit Mount Slesse at approximately the 7,600-foot level, just below the third peak. One section had plunged two thousand feet down a precipice and lay buried in snow. Another section dangled by control cables from a cliff ledge. Smears of oil streaked rock faces. Signs showed a violent explosion, but no fire. Whatever happened was sudden and catastrophic.

On receiving the discovery report, the BC government ordered a forest reserve to be placed on the mountain, guarded by the RCMP to prevent public access. Coroner Glen McDonald of Vancouver requested that at least one body be brought in as evidence. When his assistant and several mountain climbers ascended, with the intention of bringing down some remains, they were nearly killed by an avalanche. Helicopter landings were impossible because of the downdrafts, and clouds masked clear views of the area. It was decided to let the remains rest in peace on the mountain, at least until summer.

Evan J. Bullock was a pilot assigned by Okanagan Helicopters to make numerous and difficult trips to transport searchers and officials to the wreckage site, and was justifiably praised for his "fantastic flying feats" in bringing Squadron Leader George Sheahan, Commander of 121 Flight, Search and Rescue, to within ten feet of the wreck. Sheahan noted to news reporters that Bullock came so close to the mountain that the rotors of the machine were only two feet from the cliffs.

"The area is quite remote and difficult to access except by helicopter," Bullock writes to the author in December 2004, recalling that perilous flight up to the site. He also self-published a book about his flying adventures titled, *My Life with Wings and Rotors,* in which he recorded the discovery:

> The weather is usually a problem in this area also. When we arrived, the clouds covered the tops of the mountains, with brief periods of clearing, and so I had to time my close-up search with those periods. We made several passes to locate the actual point of impact with no success . . . At the very top of Mount Slesse are two sharp pinnacles, and it was near the base of these, on the western slope, that the climbers had found the piece of wreckage. There is a 'V' [between the jutting peaks], so I decided to try to fly through it beneath the clouds and as I did, we saw some wreckage. There was barely enough room to get through this opening, and with some wind activity and up-and-down drafts and low clouds over the peaks, it took all of my concen-tration . . . to land on the lower promontory . . . Some parts had remained there, some had flown over to the west side, but the fuselage and bulk of the airliner had fallen down the sheer cliff to a snow glacier some 3,000 feet below.

Upon landing at the site at the base of the glacier, Bullock observed the aftermath of such a wreck:

> The smell was bad of decomposing bodies, and there was debris all over: metal aircraft parts, tires, luggage and clothing littered the glacier. Because the glacier was very steep and slides occurred regularly, it was decided at the coroner's inquest to declare the place as a shrine for all those aboard, and no additional attempt was made to remove bodies or wreckage . . . Rumours abounded about huge sums of money on board and there were many attempts made by individuals to search for it, but the RCMP kept the place guarded and safe from molesters.

But the high snows did not melt, and the mountain retained its claim on the aircraft and the personal remains. The BC government's minister of lands and forests declared Mount Slesse to be a provincial cemetery site and public entry to its slopes was prohibited. Late that summer, eight months after the accident, a funeral party was airlifted by helicopter to a shelf three thousand feet below the crash site. There, memorial services were conducted and a wooden cross erected. The mourners had only to look across the mountain to see the crash site of the RCAF B-24 Liberator near Welch Peak. There a white wooden cross bore the names of the eleven people who'd lost their lives in that mystifying accident in June 1945. During the TCA North Star service, a CPAL aircraft circled overhead, then flew fifteen miles northwest to Mount William Knight, where the Lodestar CF-CPD had met its fate on December 20, 1942. Captain Patry cut its engines and armfuls of flowers were dropped through a hole in the aircraft floor onto this site.

In the five-year period from 1953 to 1958, TCA's aircraft made 20,460 flights across the mountains between Vancouver and Alberta, and the North Star was their only lost flight. The wooden cross was later replaced by a granite cairn on the north side of the Chilliwack River, inscribed with the names of the sixty-three people who had met their fate aboard the aircraft.

The investigation into the crash of Flight 810 was inconclusive, as were other investigations into aviation disasters on these treacherous peaks. There were no living witnesses, and no airplane parts could be brought in for reassembly and analysis, so only "probable causes" provide the conclusion to the North Star's demise. The culprit could have been a loss of power from shutting down one engine for fear of fire, or severe turbulence and icing that sent the aircraft south of Green One and brought it to a dangerously low altitude.

One possibility surfaced years later, explained in Philip Smith's fifty-year history of TCA/Air Canada, *It Seems Like Only Yesterday*. Following an investigation into the sudden drop of a Vanguard near Rocky Mountain House in 1963, it was learned that a meteorological peculiarity, known as a Bishop, standing, or lee wave, can occur in mountainous regions:

> This is a strong current of air that rushes almost vertically down the lee side of a mountain range when a rare set of conditions develops. The wind must cross the mountains at a certain angle, there must be a certain mixture of changing air pressures and temperatures, a certain combination of alternate layers of stable and unstable air and so on.[6]

When checking the Vanguard incident, investigators concluded that all the requisite conditions had existed to form a standing wave, and they then suddenly recalled the crash of the North Star on Mount Slesse. As Smith notes, "Sure enough, all the conditions for a standing wave had existed on that day, too."[7]

THE HAUNTED FLIGHT PATH

While large aircraft disappear less often than smaller ones, and passengers no longer line up at dispensing machines to purchase life insurance before boarding, the combination of unforgiving terrain and weather conditions continues to take its toll. A tally of wrecks that have occurred on the flight path from the lower mainland east over the mountains through the Hope-Princeton corridor cause many to feel that it's truly haunted. Captain Steve Eyre, a Canadian Forces air navigator stationed at the Rescue Coordination Centre in Victoria, states that this perilous path has become a collecting ground for the wreckage of "50 aircraft since 1974—it's a hot spot."

Here, a navigator must be skilled in mountain flying and also in map-reading, because one wrong turn up a box valley can mean the end of the line. Instant cloud cover or fog can shroud mountains, trees, and the horizon. One pilot likened it to being inside a white pillowcase thrown into a clothes dryer. "You lose all sense of direction, all references, and quickly lose control of the aircraft. You could be flying on upward or downward elevations, or on a sideways tilt, hard to tell. It's impossible to turn around and get out—where are the mountains? Oh my God, right here! You climb until you stall; you dive until you hit something solid; or you go into a spin and end your days on a cliff face, a rocky slope or in the 'V' of a deep valley."

Ralph Langemann, a pilot and meteorologist, agrees with this scenario. "Your worst weather will always be where the valley is at its narrowest and where the valley floor is the highest. That's just a meteorological effect because where it's narrower the wind has to blow faster to get through there, and as it blows faster the pressure drops and that makes the temperature drop. Then you're much more likely to get cloud because it's associated with the drop in temperature. Also at the highest part of the valley there's a maximum lift applied to the air blowing up that slope, and again the air is expanding and then cools. And so for those two reasons you're going to find, meteorologically, the lowest cloud will be where the valley floor is the highest and where the valley is the narrowest."

These climatic facts combine to serve serious challenges to aviators. How do you avoid getting squeezed in or pushed down in such situations? Langemann quite understands the thoughts of a pilot in such a situation: "You get down in the valley and you say, 'Oh well, if it doesn't get any worse than this I'll be okay.' But by now you're so low that you can't turn around anymore, you're committed to going ahead. The valley floor is like a "V" and you're in there. It's no longer wide enough to turn. And then you come around the corner and there's the cloud sloping down and the valley floor sloping up and that means there's no opening anymore." When asked what happens then, Langemann is blunt: "You crash," he says. "That's exactly what happens."

Langemann is not surprised that so many downed aircraft are lost, but that so many are *found*, especially in such rough geography.

> In a lot of places, like in the Rockies, the trees are fairly tall and the mountain sides are steep. And so you get into the trees and your airplane is all bent out of shape . . . And with a lot of these light planes, it has stopped going forward before it reaches the ground. The trees stop it and it falls straight down so there's not even an opening in the trees where it's gone in. And it's not recognizable as an airplane shape, it's just this mangled mass of metal, below the trees. The wings have come off and the tail's off and the whole thing is wrapped into just a ball of tinfoil. You know these things are not made of very heavy material. With a good jackknife you can cut a Cessna in half, there's no problem. We've cut guys out of them with just a Swiss Army knife. They're made of thin aluminum, thinner than your cookie sheets. And they generally paint them white, so that really blends into the snow. And so this mangled mess is lying underneath the trees and the trees have folded in above it again. I'm amazed at how many are found, to be quite honest.[8]

Pilots are often encouraged by a false hope when they glimpse sunlight on the far side of a mountain peak, but that can be Death beckoning to the other side—especially near the Hope Slide area. "You're coming from the west where the Fraser Valley narrows down and becomes really mountain-ous," Captain Eyre explained to *Vancouver Sun* reporter Brian Morton, "and you're going from easy flying to hard flying. When you're in the Fraser Valley, you're not going to run into anything. It's a big, wide-open, friendly valley. Then, as soon as you're past Hope, it's not anymore. Plus you've got the winds. Plus you've got the weather that starts in the mountains. It's very unpredictable and the weather can change from valley to valley."[9]

All such factors can increase the risk of accidents when visual flight rules just aren't enough to get an aircraft through fog-shrouded twists and turns over mountainous terrain. Clouds, foreboding and ghostly, cling to the mountainsides, trailing their foggy tailcoats and concealing downdrafts that can lower the ceiling by hundreds of feet in seconds. Perhaps a pilot could attempt to fly beneath or above the clouds, but they seem endless with no escape holes. Up above, there's ice that causes the aircraft to shake, rattle, and lose altitude. Down below, the dark trees appear, suddenly too close. Can the pilot "rest" the plane on the treetops? He's heard of that being done. But the box valley is enclosed by three rock walls, can't go up, can't go down, can't turn, too close, went too far. Dead ends. All dead ends.

And another sermon is delivered on a mountain.

Requiescat in pace.

WHERE'S JOHNNY?

Johnny Bourassa disappeared Friday, May 18, 1951, on a flight from Bathurst Inlet to Yellowknife. Several months later, the Bellanca Skyrocket (CF–EQP) he had been flying was located on the southeast shore of Wholdaia Lake, Northwest Territories, intact and in good condition. There was no sign of the pilot.

The story could end there, but it doesn't. In fact the mystery of Bourassa's disappearance comes up whenever people in the North speak of unsolved mysteries, because many of the facts simply don't fit.

THE BOURASSA LEGACY

John Maurice (Johnny) Bourassa came from a well-respected Métis family. The Bourassas had come west from Montreal to Fort Vermilion, Alberta, in the 1800s with the North West Company. From 1916 to 1935, Johnny's father, Louis Bourassa, transported the Royal Mail 275 miles up and down the Peace River between Fort Vermilion and Peace River Crossing, in summer by boat and in winter by snowshoe, with dog teams, or horses and sleigh.

King George V named Louis Bourassa a Member of the Order of the British Empire in 1935 for his historic work as a river pilot. "Unfortunately, my grandfather died in Keg River of a heart attack the very night he learned that he had received this award," says Myrna Bourassa Hargrave. "My father Johnny and his brother Arthur finished his contract hauling the mail." Other Bourassa family members claimed some thirty-three honorary medals.

Born November 26, 1915, Johnny was one of six children in the family of Louis and Maria (St. Germaine) Bourassa. He grew to be a handsome young man, just over five feet eight inches tall and well built, who worked with his father every summer as a river pilot. After completing tenth grade in Peace River Crossing (now called Peace River), Johnny found summer employment with O'Sullivan and Stigsen, River Transport and Fur Traders, as a river pilot, and winter work with the post office transporting the mail over the river ice with horse and sleigh. He consequently gained a deep love for the country, its people, and animals, especially dogs. "After Johnny had spent some time with a dog, it was his for life," affirms a Peace River resident, and many photographs exist of Johnny with "his dogs." In August 1939, Johnny married Mary Purcell. They soon had two children, daughter Myrna (born October 13, 1940) and son John (called Jack, born October 11, 1942).

The backcountry did not escape the effects of the war raging overseas and on July 7, 1941, Johnny enlisted with the RCAF. He trained at the British Commonwealth Air Training Plan's No. 4 Initial Training School, graduating on April 24, 1942, and was recommended for pilot training. He then attended No. 5 Elementary Flying Training School in Lethbridge, Alberta, graduating on July 3, 1942; and No. 15 Service Flying Training School in Claresholm, Alberta, where he was awarded the Pilot's Flying Badge and promoted to Sergeant on November 6, 1942.

That December he was sent to Halifax to await embarkation for overseas. What was a Peace River boy doing in this place? His frequent letters home to his wife and children expressed moods ranging from brave

Johnny Bourassa with baby Myrna, early 1941. (Courtesy Myrna Bourassa Hargrave)

determination to bewilderment. In one letter, he vented his frustrations and Mary penned an immediate response, excusing Johnny's frame of mind as being "fed up" with the long wait in Halifax and being so far away from home—and perhaps even a bit afraid of what lay ahead. She wrote:

> Yours will be a very great job. You will risk your life, the wholeness of your body and perhaps go through a searing experience of pain. Courage in this will help more than yourself, it will help other men too. Never a war in all history demanded so much courage, pitting the perishable body of man against formidable engines of indestructible steel. Always remember though that more indestructible than steel is the immortal soul of man.
>
> Don't be afraid. I repeat it because it is the most impor-

tant thing I have to say . . . My prayers will be with you constantly, my thoughts are always of you, your welfare, your happiness and your comfort. Never for a moment let anything make you believe that your wife, your daughter and son aren't pulling for you here at home.

We are loving you constantly,

Mary, Myrna and Jack

Johnny's card, "A Christmas Message to my Wife," sent from Halifax on December 7, 1942, contained news that he soon expected to be sent overseas.

To my beloved wife:

By the time you receive this, dear, I will be on the high seas on my way over, day by day, taking me farther away from you.

Distance does not mean anything. Our love is so great that we could be worlds and worlds apart, and it can still be felt. There is no use trying to hide the fact, Xmas day is going to be the loneliest of our lives, dear. But, be strong and courageous, as the next Xmas will find us all together. We have so much to live for, to look forward to, sweetheart. Let this be a white Xmas just for you, sweet. I will be with you in thoughts all through the day.

From your husband "Johnny"

I love you, dear.

The letters and cards from his family, filled with philosophy, encouragement, and love, fulfilled their purpose. Johnny Bourassa became a Canadian war hero. On November 6, 1943, he was promoted to Warrant Officer Class 2; on February 3, 1944, he was transferred to No. 35 Squadron of the Royal Air Force; and on March 20, 1944, to No. 635 (Pathfinders) Squadron. He received his commission to Pilot Officer on April 29, 1944, and to Flying Officer on October 1 of that year.

On October 2, 1944, John Maurice Bourassa was awarded the Distinguished Flying Cross (DFC) for completing numerous operations against the enemy, "in the course of which he invariably displayed the utmost fortitude, courage and devotion to duty." The Bar to the Distinguished Flying Cross followed on February 5, 1945, with this citation:

> This captain has carried out a large number of operations against some of the most heavily defended cities in Germany, including Berlin. He has several times been viciously attacked by fighters and has displayed exceptional skill and coolness in extricating his aircraft and crew from perilous situations. He sets his mind on the task in hand, fearlessly and with a fine offensive spirit, setting a magnificent example to his crew.

The DFC and Bar were presented in Edmonton to Flight Lieutenant John Maurice Bourassa on November 29, 1947, three years after receiving official notice of the awards.[1]

Johnny Bourassa returned to Canada, still on active duty, just before Christmas 1944. The following April he was posted to No. 165 Squadron, and to No. 168 Squadron (Transport) in May that same year, where he served in Cairo from June to December 27, 1945—in his well-known former duties as a carrier of the King's Mail, this time by air. Johnny's determination to learn and excel in his duties had brought him from "ordinary" ranking to "accelerated."

Letters received from air force mates and others he'd met in England show lasting respect for Johnny's quiet demeanor and skills, and a wish to retain his friendship although now parted by an ocean. Flight Sergeant Fletcher wrote from RAF Station Warboys, Hunts, England, "After all the reminiscing you used to do about the 'Mighty Peace' it must be good to be back again, firstly with Mary and the family, and your 'proper' climate, as you put it." Fletcher relates sad news about members of No. 635 Squadron who didn't make it, one entire crew being wiped out within five days of completing their tour of duty, and concludes, "Please express my compliments to Mary for the things she has sent over to us in the past. I'm sure everyone appreciated her cooking." Another air force friend adds a note that would have gladdened Johnny's heart: "Your dog is still at the camp, apparently well cared for and fed."

A decorated war hero, Johnny Bourassa was an expert pilot and outdoorsman who loved dogs. (Courtesy Myrna Bourassa Hargrave)

Returning to the Peace

The end of the war released thousands of young men who'd trained as pilots and aviation mechanics, and there were many competitors for jobs. Johnny was lucky. The RAF and RCAF flyers had made worldwide connections, and Johnny was offered a job as a pilot with KLM Dutch Royal Airlines based in Amsterdam, Holland. He was to report on August 30, 1946, to the company's head office in New York, all expenses paid.

But Mary did not want to go to Holland. And his parents—what would he tell them? And his children? They'd been without him for so long and now he was home. Another opportunity would come where he could use his flying skills closer to home.

He stayed and found work as a bush pilot, flying for Peace River Northern and other northern airline companies. The pay was lean, as pilots were usually paid "by the mile and by the pound." Flying small aircraft in the Far North, often overloaded and underserviced, presented nearly as many risks as his wartime endeavors. But he was with his beloved family once again.

Johnny Bourassa's valor did not end with his wartime service. On September 24, 1949, he was flying a Stinson (CF-GHC) for McAvoy Diamond Drilling & Development Co. and was scheduled to pick up three passengers. Before he left Yellowknife, fellow pilot Al Finsand came in with a Fairchild 71C (CF-ATZ). It was decided they would go north in the Fairchild and combine Johnny's scheduled pickup with some fishing at Taltheilei Narrows on the east arm of Great Slave Lake, NWT.

On takeoff from Lake Aristofats, the Fairchild stalled and hit the water with such impact that the floats ripped off and the aircraft quickly sank in seventy feet of water. The pilot and one passenger drowned. Two survived due to Bourassa's quick action. A newspaper article of unknown origin—part of a collection of newspaper clippings belonging to Myrna (Bourassa) Hargrave—praises Johnny's efforts.

One of the survivors was John Bourassa, Peace River, who was the hero of the affair in that he saved his companions from possible death in the cold waters of the lake. When the two survivors found themselves clinging to the wings of a rapidly-sinking aircraft, Bourassa spotted one of the floats that had broken off. He swam to it and brought it back to the two others-one of whom could not swim. As they attempted to paddle to shore, one of the men slipped back into the water. Johnny pulled him back onto the float ...

Bourassa knew the location of the camp they had just left and the three managed to return to it. With their matches soaked they were fortunate to find a few coals still alive in a stove they'd just dumped, and after much effort these were fanned to flame and a fire started.

The men were brought back to Yellowknife and hospitalized for a short time due to shock and exhaustion.[2]

No wonder Johnny's family worried when he flew off over uncharted wilderness, but he was a pilot, and a man of the North. He was making a living for his family, who had moved from Peace River to his base in Yellowknife so they could all be together. The risks were acknowledged, but could not be feared, and Johnny had proved calm and effective in crisis situations, both in war and peace.

RECONSTRUCTING THE CHAIN

By 1951, Johnny Bourassa had amassed over ten thousand hours in the air and was now flying for Yellowknife Airways Ltd. On May 18, 1951, Ernie Boffa, chief pilot and part-owner of the company, left the Yellowknife base in their Bellanca (CF-EQP) on a trip to Bathurst Inlet. Bourassa, transporting a group of botanists, left later in their Norseman Mark V (CF–BHV). In Bathurst, the passengers would transfer to the Bellanca to go further north with Ernie, while Johnny would return to Yellowknife in the Norseman. Neill Murphy was the engineer assigned to accompany Johnny for the round trip.

Neill had grown up with the Bourassa family in Peace River and had a wilderness background gleaned from twelve years as a trapper in the barrenlands. He'd then moved to Yellowknife to apprentice as an aviation engineer. With his combination of survival skills and mechanical experience, Neill Murphy was considered a good person to have along on a flight. "If you had Neill with you, you were assured of coming back alive," states aviation pioneer Chuck McLaren.

Murphy clearly recalls the chain of events that occurred that May 18. "I'd packed my rifle in the Norseman, and Johnny had his maps and everything. But then I had to stay in Yellowknife, Ernie Boffa had something he wanted me to do there. At Bathurst, the batteries on the Bellanca were found to be dead when they tried to start it, so Ernie took his maps out of the Bellanca and flew North with Johnny's passengers in the Norseman— along with my rifle and Johnny's maps." Johnny either forgot to transfer his maps or he assumed that Boffa's maps had been left in the Bellanca. Whatever the assumption, it would prove deadly. "Johnny Bourassa and John Norberg hand-cranked the Bellanca to get it started," Murphy continues, "with Johnny making the trip south to Yellowknife alone—but now with no rifle or maps, an inaccurate compass, and an inoperable radio because of the dead batteries and poor communications systems at that time in the North."

The flight plan from Bathurst Inlet to Yellowknife involved a stopover at Salmita Mine, which wasn't operating but was occupied by a caretaker. There he'd remove the skis and come into Yellowknife on wheels. The weather between Bathurst and Salmita was poor, so when Bourassa was late arriving in Yellowknife it was thought he'd set down somewhere, likely

Neill Murphy's combination of survival and mechanical experience made him a good person to have along on a flight. (Photo by Shirlee Matheson, August 2003)

after detouring off course. Perhaps he hadn't been able to get the engine restarted, as he'd have to swing the prop on his own; if the battery was still dead, the radio wouldn't work either.

It was not until May 20 that Ernie Boffa, after returning to Yellowknife from his trip to Spence Bay, advised Yellowknife Airways' office in Edmonton that Bourassa had not reported in for his planned stopover at Salmita Mine. That evening, Boffa commenced his search for Bourassa between there and Bathurst. When nothing was found, "Westy" Westergaard, the Edmonton manager, called the search and rescue unit for assistance. "'Westy' was wringing his hands for two days, but he wouldn't call the air force in for a search," Neill Murphy says. "He had to wait until Boffa came in, as he was one of the owners. We were all pretty worried about it. I said,

'To hell with Boffa. Call the air force in anyway! Tell them one of our aircraft is lost.' Finally Westy did. The air force was in the next day of course."

A RCAF Dakota (C-47) left immediately from Cambridge Bay, flying Visual Flight Rules contact with the ground to make detailed searches, while Boffa searched with the Norseman in the same area. When nothing was found, the air force mounted a full search involving five Dakotas from Edmonton, two Canso Flying Boats from Vancouver, and a Lancaster from Fort Nelson, along with observers from No. 7 Company, Canadian Forest Rangers. But poor weather—with rain, wet snow, fog, and low ground ceiling—caused delays, and ten days passed before grids could be established between Bathurst and Yellowknife for low-level searches.

Neill Murphy was brought in to discuss the search pattern:

With six hours fuel and flying at 115–120 miles per hour, I measured out on a map the circumference to the east and west, in a *southerly* direction, that they should search. But Boffa told them, "He's got to be between Bathurst and Yellowknife, somewhere in those lakes." I don't know why he thought that, but he insisted.

He also said, "If you're looking for a crashed aircraft, you're looking for a burned aircraft and a dead man." That gave a false impression of what might have happened. The air force is often criticized for not listening to the bush pilots, but they listened to Boffa who was considered a bush pilot, and that's why they just searched in a northerly direction from Yellowknife. That's the way Johnny was supposed to come, so they felt it was logical to search that area.

But what made me so mad, and so many other people mad too, is that's where their mental block came in. They would not consider that he might be alive, and have landed somewhere else. I told them, when the search was first on, what Johnny always said: if he was lost he'd pick a direction and follow it in order to find something familiar. That's exactly what he did. But they wouldn't listen.

By June 17, when the search was called off, thirteen aircraft had searched 160,000 square miles of territory.

In an interview with the author, Neill Murphy takes a string, holds it at Bathurst, and swirls it down and around over the map. "If they had taken this triangle—in a line from Bathurst southwest to the southeast with Yellowknife in the middle of the swing—they'd have found him. If Johnny heard them he could have lit a fire, even if he had to burn the aircraft! But there was lots of wood. Sure, it would be expensive to search a 90-degree sweep flying at 1000 feet in grids, but I begged them to at least make a couple of passes. But they just searched straight from Bathurst to Yellowknife." With painful pride in his voice, Neill Murphy says, "I was right."

Meanwhile, Mary Bourassa and the two children, Jack, now eight years old, and Myrna, ten, were existing day by day, praying for positive news. Oh, where was he? Surely he'd make it home! Hadn't he once carried the mail many many miles over untamed wilderness, from Peace River Crossing to Fort Vermilion? He knew rivers; he knew the woods and the animals. He could tame the wildest dog into loving submission. He wasn't afraid of anything. But *they* were.

Myrna clung to images of her dad's heroic war action. She thought of him returning home from overseas, a beloved stranger whom she regarded with a sense of awe. Sometimes he would take her on his knee and she'd be so filled with emotion at his being home again, and her being so close to him, that she'd burst into tears.

He'd shown tears too, twice that she could recall. When Myrna's maternal grandmother died, her father was the one who broke the news to the children, and he'd had tears in his eyes. He also cried when he had to end the life of their Airedale dog, Joe, who'd been so protective that one day he'd bitten a child he thought was harming them. Johnny loved dogs so much, and Joe was like one of the family. Everyone had cried then.

Myrna had kept the letters written to and from her dad when he was away at war, and from when he was based in Yellowknife with the family still in Peace River. She'd write how she missed him, and back would come his letter with her allowance tucked inside.

She'd been happy, and a bit scared too, when their family moved north to Yellowknife. One night she and her dad had walked home together through the cold, clear air. He'd held her hand and talked about flying, and had all sorts of stories. Now the stories were about him. In class, a boy stood up and related the news of Johnny Bourassa's disappearance for his current events assignment. Myrna had sat in silence, eyes lowered, staring down at her desk. No teacher, or anyone really, had talked to her about what was happening.

A local paper, the *Yellowknife Blade* ("The Only Independent Newspaper in a Million Square Miles") offered encouraging reports.

> When he was a Pathfinder overseas, Johnny built up a reputation for efficient and cool-headed flying and is withal a good bushman. There is no man in the North more able to take care of himself in an emergency.[3]

Letters flooded in to Mary and the children from friends and relatives. One letter was especially poignant, from Mabel Denison, the widow of pilot Dick Denison, who had flown an Anson (CF-EEJ) for Yellowknife Airways.

> I was terribly grieved for you when I heard over the radio that your husband is missing. I do hope you have better news of him by now. Was he flying in the same outfit that Dick was in? People have been so good at helping us to recover Dick's things. I can only hope that you won't have to go through the heart-

break that I have had to go through. That Northern flying is hard on the pilots but they all seem to like it. With best wishes for better news . . .

On February 6, 1951, Dick had taken off on skis from the old CPA base in Peace River and crashed into a nearby house, instantly killing himself and passenger Earl Bulmer. Myrna could recall her dad hauling tackle to drag the river in search of this friend who'd crashed his plane near the bridge and drowned. He did find a body, but it was of a woman who had jumped into the river up by the hospital and drowned. It was the first time Myrna had heard the word "suicide."

Two weeks passed, then three, with no sighting of Johnny's aircraft. Letters and calls of encouragement continued to arrive, but hope faded as time passed.

> My dearest Mary and Children [wrote Johnny's mother]:
>
> I know how you feel, my dear. We must try and be brave and hope for the best. My dear son Johnny, [in] whom I always had so much faith and confidence—still we never know. I only hope he is all right, as you know yourself you have all the trust in his work, but dangerous. I try and be brave and pray . . . Somehow I can just picture him sitting by a campfire and waiting for someone to assist him when the weather clears so they go on for the search.
>
> How is my dear Myrna Jean and Jackie? How are they taking it? I hope they will understand about their dad not returning. I just hope that he comes with his plane, if possible.

Johnny's sister Florence and her husband, Steve Pim, offered encouragement and strong advice from Guelph, Ontario, in a letter dated May 29, 1951.

We have the greatest faith in Johnnie's ability—he is so resourceful, there is no one like him. Steve is so encouraging, he thinks there is no one like Johnny, and today said that when he gets out of this scrape we must do all we can to get you all to move down east, away from that frightful country. He can get something to do here and take it easy for a change. So you must really do all you can to make him stop 'bush flying,' Mary, it's too nerve-wracking.

SEARCHING FOR JOHNNY

On August 31, 1951, Labour Day weekend, a USAF B-17 pilot supposedly searching for a lost aircraft told Art Spooner, the operator at the Cooking Lake floatplane base, that he'd spotted a downed aircraft near Snowbird Lake. It was initially described as "a silver Norseman on skis, red wing-tipped, letters CF-EQP, on the beach, no people." Tommy (T. P.) Fox, who had recently acquired Yellowknife Airways, immediately recognized the registration. He dispatched the company's Fort Smith–based pilot, Gordon Cameron, to fly to the site with an RCMP officer. Cameron's telegram confirmed it was indeed Bourassa's airplane. Fox relayed the information to the Department of Transport.

The aircraft, spotted by the USAF 118 days after its disappearance, was nestled among shoreline boulders on the south side of Wholdaia Lake, NWT, 325 miles southeast of Yellowknife. The starboard leg was partially torn from the frame and one prop tip was bent. Johnny Bourassa's last communication was found inside the cabin.

Departed from here Wednesday, May 23, 1430 hours (2:30 PM). To whom it may concern.

I have gone to the southwest of this lake and from there

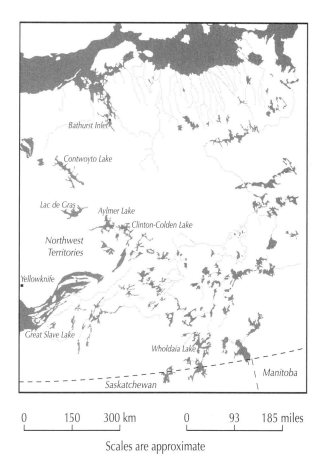

| 0 | 150 | 300 km | | 0 | 93 | 185 miles |

Scales are approximate

I intend to walk a northwesterly direction. I believe that it will take me to Great Slave Lake, somewhere near Reliance.

I have a good two weeks' ration that could be stretched to three weeks. I hope to be somewhere before then.

Please be on the lookout for a smoke signal as I will endeavor to flag any aircraft down that should be flying overhead. A lift will be appreciated.

Why this aircraft happens to be sitting here: took off from Bathurst for Yellowknife without any maps and no timepiece. The clock on the instrument panel would not function (I have made it function since, as I have it with me now). I flew under overcast for some time, holding 180-185 magnetic on the compass (I found the compass very inconsistent). Then I broke out into sunshine, scattered cumulus clouds and rain showers.

I mis-identified Aylmer and Clinton-Colden, took them for Contwoyto and de Gras [lakes].

I must have passed well east of the east end of Great Slave Lake. By the time I realized my error I was too low in gas to do much but find a suitable landing.

I fractured the starboard leg when beaching the aircraft. I was traveling at a fast clip traversing bad ice and she swung just as I cut the engine. She straddled the boulder. I meant her to be on the left hand side. There was quite a cross-breeze blowing.

I am leaving my bedroll behind. Please take it back to Yellowknife as I have some personal stuff wrapped in it. In the meantime I hope to be either in Reliance or Stark Lake before long. So sorry this happened.

Johnny Bourassa

The date on the note, May 23, indicated that Bourassa had spent five days beside the slightly damaged Bellanca before leaving the site. The *Edmonton Journal* mourned the unexplainable disappearance of "the famed bush pilot, who won the DFC [and bar] during the Second World War . . . and who became Canada's prime mystery Tuesday when his plane was found."[4]

A northern weekly geared to industries of the Pacific Northwest, *The Machinery Record*, offered a practical analysis of the lost pilot's skills and his ability to survive.

Johnny stems from Northern stock. His grandfather and father ran the Royal Mail by dog team in the early days, so Johnny knows the North. And the North knows Johnny; knows that he is strong and vital and not an easy victim to her ways.

Johnny has called on the lonely trappers on the Mackenzie, prospectors on the barren lands, delivered food and essential supplies on the Athabasca, flown the sick from the Peace, the Great Slave and the Lesser Slave—Johnny who never spoke of himself, but who held the deep respect of all of us who knew and worked with him; Johnny, emblematic of all our bush pilots, our air ambulances, purveyors to the North, to coastal camps—typical of the romance of flying and hero to the hinterland. Will he walk in?[5]

When the RCMP detachment in Fort Smith reported on the long, unsuccessful searches made by the RCAF Air Rescue Squadron and ground search parties, and accidental discovery by the USAF, officials were puzzled. Back and forth went letters of admonishment and explanation among the Department of Transport, Edmonton air traffic control center, and the district controller regarding search directives.

The new search master was Wing Commander C. M. Black from the RCAF station at Churchill, Manitoba, assisted by Flying Officer W. H. Wilson of the rescue coordination staff in Edmonton. More military search and rescue aircraft were brought in from Winnipeg and Rivers, Manitoba, and from Edmonton and Vancouver, as well as civilian aircraft and observers. The air force set up camp on a sandy spit on Woodruff Lake to bring their search planes into the area and scrutinize a fifty-mile-wide strip over the 240 air miles between Wholdaia Lake and Fort Reliance.

Flying Officer Clair Gleddie and Flying Officer Martin "Mo" Aller recall flying a Dakota off Yellowknife's gravel runways from September 9 to19, 1951, in search patterns at five hundred feet above ground level, with the lines two miles apart. Below them, herds of caribou were spotted, with wolves hot on their trail. At least there was food available, they thought, so Bourassa wouldn't starve if he was uninjured and able to bring down an animal.

The Bellanca at Wholdaia Lake. (Courtesy Myrna Bourassa Hargrave)

Northern ground crew, including Neill Murphy, were again called in to help. With his wilderness knowledge, as well as his association with Johnny Bourassa, Murphy set out to find him. It was known that Bourassa had a ration kit, an ax, fishing gear, and a bedroll—which, as he stated, he'd left in the airplane. It was initially believed that he had a rifle, but in actuality the rifle had belonged to Neill Murphy and had stayed in the Norseman. Neill knew that Johnny never wore a watch but had fixed the clock, which could work as a sundial to help him determine his position. The stove (blow-pot) would have been too heavy to carry, as would the canister-type flares.

Murphy explained his theory to the searchers: "Johnny picked a direction in the hopes of seeing something familiar . . . but he just went too far east. He got lost, confused. He ran short of gas before he could get oriented. He just isn't familiar with this part of the country. At the time he landed the aircraft the ice was strong enough to hold it until he was on the beach—his note states that he came in fast because of the poor ice. But after five days he probably thought it more sensible to walk around the bay."

The air force paired Murphy with a young recruit named Kruger, so he could gain some wilderness experience, and together they surveyed the area around the abandoned Bellanca. They saw no sign of any trail, although some reports later said there *were* visible trails heading east in the direction of Stony Rapids—the opposite way to Bourassa's planned hike. The first item spotted was a broken dog collar and a couple of straps hanging on a bush near the airplane. The weathered harness appeared to have been there for some time. "Maybe Johnny saw this and it gave him the idea he was closer to a settlement, or to Reliance, than he actually was," Neill mused.

Murphy and Kruger, walking from the abandoned aircraft to the left (west) in the direction that Johnny's note indicated, arrived at the southwest bay of the lake. There they found evidence of his first camp. Murphy says, "First, we found a chip of wood from where Bourassa had made a shelter. The camp we found was his, no doubt, because of a used tea bag and wrappers from the high energy candy that was packed in the ration kit. Next, we found a sapling with the identical cutting to the wood chip back from his first camp. I showed this young air force fellow where the sapling had been cut down and used to help him get across a creek—it would have been higher water back when he'd crossed it—and then he'd discarded it on the far side. So we had proof that he'd got that far."

Finally, a few miles northwest of the Dubawnt River, which flows out of Sandy Lake, they found another clue. "A log had washed up on shore, all trimmed, that he could have used as a raft," Murphy says, "but we found nothing further. And there was no evidence that this log had been used by Bourassa. Somebody had used it, but there was no sign of who it might have been. That was the end of any evidence we saw that Johnny Bourassa might still be alive." Murphy had personally covered 150 miles during his ground search.

Northern bush pilot Jim McAvoy, who joined the civilian pilots on the search, agrees with Murphy's ideas of what likely happened to Johnny Bourassa. "He just got lost," McAvoy recalls. "He was not much of a navigator. He always had a tendency to go to the left. Johnny could have fallen through the ice. Indians could have done him in. They were an isolated bunch up there, and they'd been watching him from across the lake for five days! When I came down in the aircraft and tried to approach them, they ran into the bush, as if they had something to hide. This was highly unusual—most times Native people, or anyone living in the bush, when they see an aircraft come down to greet it."

Neill Murphy doubts that the Native people would have harmed Johnny, although he acknowledges that there might be a reason they were reluctant to investigate, or come to the aid of the pilot. "The game wardens had visited these people that very day, and Johnny landed there in the afternoon. When they saw the second airplane come in they could have assumed it was the game warden, so they just sat and watched him. That's what I understand. They might not have been able to see him, but they could see the airplane. They never went across to check him out, but by that time the lake was no doubt filled with rotten ice. In order to get over to him they'd have had to go around the end of the lake and back, a distance of anywhere from three to five miles." If Bourassa was aware of the Native camp, it's thought he would have mentioned it in his note. "Three or four miles is a long way to hear dogs barking," Murphy says. "They would no doubt try to keep the dogs quiet in hopes of caribou showing up anyway."

Near the end of the prescribed search, Neill Murphy was flying with Flight Lieutenant Carroll Potter, who assisted with the air force Norseman to "jump" the ground searchers over lakes or rivers. At one of the last places searched by ground they saw three burned standing spruce trees. The fires appeared to be a couple of months old and deliberately set under three different spruce trees with no connecting fire marks. Could they represent a "three shots" distress signal? They were puzzled. Why hadn't Johnny left any tree blazes to assist searchers and also to ensure that he wasn't circling on his own trail? His trail did not show signs of a seasoned northern bushman, and so the rumors began, some wild and ridiculous, others hurtful, but none conclusive.

Yellowknife's *News of the North* met the allegations head-on in an article published September 21, 1951. Base camps for the ground search parties were being closed and soon neither float- nor ski-equipped aircraft would be able to land on the small northern lakes as they started to freeze, but this did not mean that the search from the air would be discontinued. "This is only one of the many wild rumours which have gained wide circulation during the period since the search was resumed," the article went on to say. "Among others were stories that Bourassa had been found—and many of the tales were told with a ring of authenticity—which was entirely deceiving. Many citizens are perturbed and feel the inventors of such stories should be punished."

The second search formally ended on September 27. An area of fifty-seven thousand square miles around Wholdaia Lake had been carefully checked, with nothing to show of Bourassa except signs from the first two camps and the three burned standing trees. The RCMP filed their report on September 30, 1951. "The old camping place that was believed to have been used by Bourassa was found approximately 50 miles northwest of where the aircraft was found, and in the direction of Reliance." That was it.

Civilian pilots continued to search on their own until the end of November, and watched for any sign of Bourassa in the Reliance to Wholdaia Lake territory. The Bellanca remained on the beach, forlorn and forgotten, as it awaited the return of its pilot.

The RCMP kept the file open, even though there seemed little likelihood of finding Johnny Bourassa alive. In May 1952, the Order of Presumption of Death was authorized, "due to exposure or drowning," and a death certificate was issued on December 15, 1952. It was Mary's sad duty to inform family and friends that it was all over. Now, she and her children had to pick up the pieces of their lives without their husband and father.

Equally heartbreaking were letters that continued to arrive from Johnny's former RAF mates who had not heard of the tragedy. Norman Rowell, Sunderland, County Durham, England, writes of taking his four-year-old son to Elvington in Yorkshire to see the old station, and named some of "our pals":

> Well, the old place is still there just as it used to be—except it's a 'ghost' station now. The RAF have abandoned it and the weeds have grown wild. I paid a visit to the old Sergeant's Mess—did that bring back memories. I stood there for a full 20 minutes after having fought my way in through the weeds around the door. The voices I heard all over again—my eyes filled with tear[s]—and my heart was full.
>
> Will you ever forget, John—of course we shan't—I went into the old 'briefing room'—I could see us all sitting there—we were all together again. Sentimental yes—but money could not buy—nor shall time ever steal away—those memories of the grandest bunch of fellows I ever met.
>
> I still treasure your friendship John—so write, old boy, when you have a spare minute. Give my regards to Mary and your children, John.
>
> All best wishes, your sincere friend, Norman

By the time the letter was received in Peace River, Johnny had been gone for over a year.

"I remember leaving Yellowknife and traveling on the Beech Expediter," Myrna says, recalling that painful period when the little family of three packed their belongings and returned to Peace River. "I don't remember ever being happy. I still am trying to understand all the ways that his loss has affected my life."

The Bellanca was finally dismantled in March 1952, flown to Edmonton in a Bristol Freighter, and sold to Superior Airways Limited of Fort William (now Thunder Bay), Ontario. Then, on September 30, 1964, it was destroyed by fire following a forced landing at Horsetail Lake, Manitoba. All was gone, including its log books and certificate of registration. "EQP" disappeared as completely as had its once-famous pilot.

DISAPPEARANCE THEORIES

Reports of missing airplanes in the vast territory and volatile weather conditions of Canada's North were not uncommon. But in this case, speculation began that brought the Bourassa story out of the ordinary. The most practical responses were issued by on-the-spot northern colleagues, such as Neill Murphy, who stated:

> Where the aircraft went down, the closest place is Stony Rapids. People forget, when they make assumptions that there is some mystery here, saying Johnny staged his own disappearance, that a stranger doesn't just walk in to small isolated communities surrounding the area where the aircraft went down—such as Stony Rapids, Fort Smith, Roche River or Snowdrift—without being observed.
>
> Secondly, these communities are hardly within walking distance. The air mileage from Bourassa's aircraft on Wholdaia Lake to these various sites, which does not account for obstacles such as rivers, lakes, or hills that a pedestrian would encounter, is: 240 miles to Fort Reliance; 272 miles to Fort Smith; 304 miles to Roche River; 145 miles to Stony Rapids; 256 miles to Snowdrift. Impossible!
>
> Johnny just got lost," Murphy concludes. "People credit

him with more wilderness survival knowledge than he actually had. His father, Louis, ran the mail and was a river pilot, and Johnny also piloted boats on the river. But they weren't *bush* people. He never trapped.

The pairs of lakes that Bourassa had mistaken, Aylmer and Clinton-Colden for Contwoyto and Lac de Gras, are separated by one hundred miles. While their shapes might be similar, they are known landmarks to most "north of 60" pilots. Furthermore, if the fuel indicator was working on the Bellanca—even though the compass and clock were not—he surely would have noted the fuel level going down and realized the difference between three and six hours' flying.

Jim McAvoy agrees with Neill Murphy's assessment of Bourassa's navigational and wilderness skills. "He wasn't too good as a bush navigator—or an aviation navigator. Even if he'd had maps with him, maybe he didn't know how to read them. In the air force he was used to having a navigator with him to tell him where he was going." Bourassa had flown for McAvoy Diamond Drilling & Development Co. before signing on with Yellowknife Airways. "One day I was flying with Johnny and he couldn't figure where he was," McAvoy continues. "There was smoke from Giant Yellowknife Mine which was a good indicator, but Johnny didn't recognize it; instead he went east to the shore of Great Slave Lake."

The generators in the Bellanca would likely have cut in at 1,100 to 1,200 rpm to activate the radio—that is if the radio was working at all. "Radio communication in the North at that time was not much good," says McAvoy, "and anyway, it wouldn't have made much difference because Johnny didn't know his location to be able to tell anyone."

Another theory—one which his family prefers—is that Bourassa fell through the ice. But Neill Murphy's view is that "if Johnny had wanted to cross the ice he would have crossed from where the aircraft rested instead of going to the end of the bay where we found evidence of his camp. But the ice would have been so bad at that time of year that he wouldn't have risked crossing. He could have encountered a bigger lake, but I don't think so, although there are creeks and rivers all over the place." Murphy continues, "I think what happened is he started to walk and that was it. Ran out of food. Maybe the bugs drove him crazy. It was the eighteenth of May when he went down. It wouldn't have been too long before the mosquitoes were out, and right after the mosquitoes come the blackflies, which are actually worse."

There have been a number of cases where people lost in the wilds have been driven mad by insects—mosquitoes, blackflies, sandflies, "bulldogs" or horseflies. Building a smudge from green wood might help, but did he have matches? In his note, Bourassa had asked searchers to watch for a smoke signal. Why was his smoke never seen by searchers?

Murphy notes that black bears are the only animal in that area that could have caused him any trouble, and even they wouldn't be a problem unless he ran into a sow and her cubs. Even so, Murphy considers that possibility to be remote. "I don't think the barrenlands grizzly would get down that far because that area is all bush. There was no danger from wolves—and they've never killed a human anyway." Murphy says, "He was low on rations, had no gun, and didn't know where he was. It was just an idea he had of heading up toward Reliance. It's hard to say what goes through a person's mind when they're lost."

Could Neill Murphy, a wilderness man himself, have stayed alive under such conditions?

> To start with, I wouldn't have left the aircraft. There was fishing tackle in the aircraft, hooks and line—no pole or reel—just basic stuff, and he was there at a lake. There was bound to be fish in it, or in a nearby creek. He had a couple of cooking pots. I think I could have survived, yes, by staying there and fishing. He also had snare wire. I imagine he knew how to set rabbit

snares, but as far as a lot of wilderness experience, I don't think Johnny had it.

I don't know whether he'd eaten all the rations because he was there five days, but in his note he said he had two weeks, plus another if he stretched it a bit. There may have been last year's berries in some places, cranberries, kinnikinick, stuff like that. So, he had wood, he could make a fire and keep warm. Had he stayed with the plane he'd have been alive. But after five days I guess you'd get thinking that nobody is coming to get you so maybe you should just walk. And notoriety comes in there too, you know, and pride, of walking out on your own.

Bourassa's clothing for that trip would also be cause for concern. He was known to wear leather "oxford" shoes inside buckled rubber overshoes—hardly hiking gear. On the positive side, he had taken the clock from the airplane to help with direction. (He took "sun-shots." By pointing the hour hand on the sun, he would know that midway between 12:00 and the hour hand would be south.) He also wore a parka and took the ration kit.

Neill Murphy is not surprised that Bourassa's body was never found. "Bones would be all scattered about." But the ax, ration kit, clock, clothing items? Not a sign. That brings the "new life" theorists to the fore, with compelling viewpoints that continue to cast a mantle of mystery over what might have happened to Johnny Bourassa. Carl Huff from Westlock, Alberta professes a long-time interest in the Bourassa case, and has compiled notes over the years from reported statistics and interviews. Although Johnny's family and many friends do not entertain the possibility that he disappeared of his own volition, Huff believes it is entirely plausible. "There is a mystery here. Many things just don't fit," he says, and lists them.

- Johnny *knew* where he was. His note said so. He planned to walk to the eastern shore of Great Slave Lake, which he thought might take thirty days.

- He was inexplicably far off course [approximately 45°].
- He spoke English, but Huff believes he also spoke Cree and French. (Murphy refutes this point. "He may have had a smattering of both but by no means enough to talk anyone into helping him skip the country." Myrna (Bourassa) Hargrave agrees. "My paternal grandmother spoke both Cree and French as she worked as a police matron and was often called upon to use these languages in the course of her work. But I never heard dad speak any language other than English. If he did, it was very little.")
- Some reports say that his bush skills were extraordinary. He could handle boats, dogs and horses, swim and fly airplanes. Yet he left no trail, such as tree blazes, for searchers to follow. Why not?
- Across the lake from where he'd beached the aircraft, three or four miles distant, a group of Natives watched him during the five days he spent with the aircraft. Surely he'd have heard their dogs barking and gone over there!

These and other factors cause Carl Huff—and some others—to feel it was quite possible that Johnny staged his own disappearance. "He could have had money on him if he'd come across the lost US mail plane which had crashed off-route, and found cash aboard," Huff says.

People who knew Bourassa take issue with Huff's theories. "What mail plane?" Neill Murphy asks "The USAF B-17 was coming down from their big air force base at Thule [Greenland] by Melville Bay to Edmonton, the control district for this part of the country, which is in a straight line that passes close to Wholdaia Lake. So maybe the US air force wasn't even searching for a downed airplane."

Carl Huff continues with his rationale: "He could have made it to Fond du Lac or Lac La Ronge and, speaking Cree and passing as a Native as he was Métis, hitched a flight out and away."

"That's impossible!" Neill Murphy says, "People would have recalled this happening and reported it. And it would be impossible to pay for a flight in American cash and not be noticed, especially in the ragged condition he'd have been in after many days of wilderness survival." Murphy adds, "There was also a lot of talk about money already being on board Bourassa's airplane, but it wasn't true. The Hudson's Bay post—where he was coming from at Bathurst—never carried any money; it was all trade."

Again, Jim McAvoy agrees with Murphy's assessment. "No way could or would Johnny have staged his own disappearance! He couldn't have existed there on his own, or get out without help. He never knew where he was anyway. If he had, he would have gone to Fond du Lac, much closer than Reliance. [He] didn't have skills to 'stage' a disappearance. If he'd had such skills he wouldn't have got lost in the first place."

Myrna (Bourassa) Hargrave remembers riding with her father in a car, so many years ago. "My father seemed to drive in the middle of the road. That was the joke in the family at the time and after, that he drove a car like he was flying an airplane, as if he had the whole road to himself." She smiles sadly at the recollection. "Given his car driving habits, perhaps he was a bit of an underachiever regarding navigation."

The stories have doggedly refused to die. "There were a lot of rumors floating around at the time, and there are still a lot of persons here who 'knew of a friend' or who 'has a friend' who saw Johnny walking the streets of London, etc.," Neill Murphy admits. "I would imagine that 99 percent of the men who went overseas either had a girlfriend or had casual acquaintance with one. Some may have been more serious than others. Maybe his was, I don't know. Don't forget that he had a wife and a couple of kids, he had a job here, so personally I think that when he left England he forgot about it—if any love affair existed there to start with."

"CANADA'S PRIME MYSTERY"

The name of Johnny Bourassa arises whenever mystery surrounds disappearances in the North. When Blake MacKenzie left a note in his downed aircraft in the Nahanni region and then disappeared forever, the *Kelowna Daily Courier* headlines on August 16, 1962, read "Lonely North Still Grips Secret of Missing Pilots."

> The silent lonely north, and a message found in an empty damaged plane today, clad in mystery the fate of a pilot who disappeared months ago—just as it did 11 years ago. The rugged wilderness offered few clues to what happened to Edmonton airman Blake MacKenzie. The disappearance roused echoes of the dramatic 1951 search for northern bush pilot Johnny Bourassa.

Both pilots disappeared on flights between remote northern points; both violated a cardinal rule of survival by leaving the site; discoveries of their messages came months after searches had discontinued for their missing single-engine aircraft. Searches unsuccessfully covered 118,000 square miles for MacKenzie, and 160,000 square miles for Bourassa, until both planes were found by accident long after the formal searches had been abandoned.

"Today," continued the *Kelowna Daily Courier*, "in a rugged area far to the west of where Johnny Bourassa disappeared, searchers were to hunt clues to the path followed by Blake MacKenzie." Neither pilot was ever found.

THE SAGLEK SAGA

Saglek Fjord drains Labrador's inland rivers to the northeast, then empties its ice-crusted waters into the Atlantic Ocean. This lonely coastline is home to wild animals, such as foxes, wolves, and seals, and the few people who've learned to survive on its cold barren land. The area has also become an important site for North America's air defense—and was the setting for a chilling World War II drama.

The B-26 Martin Marauder bomber, developed by the Glenn L. Martin Company in Baltimore, was a tough plane to fly. The twin-engined aircraft was rugged and heavily armed, but hot and fast, and the high accident rate soon earned it the nicknames "Widow-Maker" and "The Martin Murder." With time, however, the B-26 became an effective wartime bomber.

During World War II, artists often decorated combat aircraft, especially American, with cartoon artwork. This "nose art" might feature pin-up queens, or caricatures of fearless heroes, or animals uttering war cries. These images brought recognition, pride, and often legendary status to the artists, aircraft, and crew.

In the autumn of 1942, aircraft and personnel were desperately needed to support the North African Campaign. The United States Army Air Corps

440 Bombardment Squadron, 319 Bombardment Group, commissioned Martin Marauder B-26B (Serial Number B-26B-2-MA, NO.41-17862) to join the fray. The nose art on this Marauder depicted Snuffy Smith, created by artist Billy DeBeck for his popular *Barney Google* comic strip, complete with helmet, rifle, and bayonet, bravely charging forth, urging action: "Time's a' Wastin'!" For this mission, the seven-member crew aboard the Marauder were:

- Pilot, 1st Lieutenant Grover Cleveland Hodge, Biloxi, Mississippi;
- Co-pilot, 2nd Lieutenant Paul W. Jannsen, Worthington, Minnesota;
- Navigator/ Bombardier, 2nd Lieutenant Emmanuel J. Josephson, Port Richmond, Staten Island, New York;
- Radio Operator, Corporal Frank J. Galm—later corrected to Galms—the Bronx, New York City;
- Gunner Corporal James J. Mangini—spelled various ways—Lyndhurst, New Jersey;
- Engineer, Sergeant Russell Weyrauch—later corrected to Reyrauch—St. Joseph, Michigan; and
- Passenger Technical Sergeant Charles F. Nolan, New York City.

It was an exciting time in the young men's lives, and First Lieutenant Grover Hodge, called "G" by his family and close friends and "GC" by others, planned to record his experiences in a journal. On October 13, 1942, he piloted the Marauder from the USAAF North Atlantic Wing, Air Transport Command base at Presque Isle, Maine, to Goose Bay, Labrador, "over some of the wildest country I have ever seen—the Canadian Wilds."

Rough weather held back flights, and the Marauder's crew was forced to "waste time" in Goose Bay before flying on to the American air base of Bluie West-1 (BW-1) at Narsarsuaq, Greenland, on October 21. From there they planned to proceed to Reykjavik, Iceland, and to Prestwick, Scotland, to join the action in North Africa. The weather worsened, with snow, rain, and wind gusts up to 112 mph. After sitting around BW-1 for a month, patience was wearing thin. Flying was definitely out, but "G" Hodge and his crew remained enthusiastic about carrying on their work, although like the others, Hodge missed his family—and his fiancée, Velma Fern Carter.

Finally, on December 8 they were given clearance to leave BW-1, but thick clouds forced them to return to base. There was no question of continuing on to the combat zone. They'd return through Goose Bay to Presque Isle and reroute through the South Atlantic. Clearance was given the morning of December 10, 1942. *Time's a' Wastin'* took the lead position in a flight of four B-26s from BW-1 to Goose Bay Air Base in Labrador, and their fate was sealed.

First Lieutenant Hodge's journal entries reveal details of what happened next. (The following portions of the journal are quoted from the transcripts received by historian Clarence Simonsen from HQ, USAF Historical Research Center, Maxwell AFB, Alabama.)

10 December, 1942

Took off at last for Goose Bay. About 1 hr. & 15 minutes out we ran into some clouds and I turned around and called for the formation to turn around also. One ship came out (I think I saw the other two later) but I lost him while getting down below the clouds. We found an opening to the SW at about 2,000 ft. and after flying in that direction we broke out. We finally had to go back up to 13,000, but it was clear sailing, so we kept on.

Lt. Josephson gave me a heading to get back on course, but we know now that it was too big a correction. About halfway over I picked up Goose Bay beam, but the set went dead after a few minutes. It was too late to turn back then so we kept trying to get it on the compass & liaison sets but couldn't.

We finally hit the coast and decided that we were south

The crew in front of the Martin B-26 Marauder, Time's A Wastin' (clockwise from top left), 1st Lt. Grover Cleveland Hodge (pilot); 2nd Lt. Paul W. Jannsen (co-pilot); Cpl. Frank J. Galms (radio operator); and Bailey (first name unknown) (engineer). (Courtesy Alma Rose)

of G. B., so we turned North. When we finally realized we were [too far] North we were almost out of gas. As we turned around I started looking for a place to land. I wanted to get back to where there were trees but the engines started missing so we came on down. The crew never batted an eye when they were told that we were going to make a crash landing.

Even if I do say so myself it was a good landing—and Lt. Janssen did a good job of cutting the switches. We hit a rock that tore the bomb bay open and one prop tip went through the fuselage right behind me, but outside of that the ship was intact. It swung around almost 90° before stopping; the fuselage made a good wind brake that way. (Hit at 1855 GMT) It was almost dark so after eating a cold ration we went to bed in the ship. We had 17 blankets, a comforter, and a bed roll but we were still cold—and couldn't stretch out either. None of us slept very well.

I spent the night figuring out where we were. Lt. Jos. took a star shot and next morning we compared notes and decided that we were about 300 miles north of G. B. at about 58° Lat.

The wheels-up landing was hard, but the aircraft sustained minimal damage, coming to rest at the base of an eighteen-hundred-foot hill. With the health of all crew members ascertained, they began to make survival plans pending rescue. Hopefully the other three B-26s had landed as scheduled at Goose Bay. They would report them missing, and a search and rescue party would be dispatched.

11 December, 1942

Lt. Jos. walked to the fjord to the west and Galm to the one on the east. We spent most of the day cleaning up the ship and pooling all the food. In the afternoon Lt. Jans. and I climbed the mountain in front of us but didn't learn much. Nolan worked on the put-put all day without results. Late in the afternoon we made a fire out of gas and oil in a tin can. That night, as on the 10th, we cranked the dinghy radio. It was pretty windy, so we spent the night in the ship but still didn't do much sleeping.

Their auxiliary power plant, called a "put-put," was of vital importance.

Clarence Simonsen's painting of the crash site, December 10, 1942. (Courtesy Clarence Simonsen)

Stowed in the waist compartment of the aircraft, it provided an output of seventy amperes to boost the electrical energy for starting the engines. If the put-put was inoperable, a signal could not be sent for help. But other matters were of equal importance—fresh water, food, and heat.

12 December, 1942

Made 3 big improvements in our situation. Lt. Jans and Galm discovered a lake close to the ship and saw a fox, Weyrauch and I saw 50 seals in the fjord and we saw wolf or fox tracks, so we know that there is food here. Also we made a lean-to out of tarps under the wing on the lee of the ship and we slept out there. It was much better, but still not very satisfactory. Before going to bed, we cranked the radio again and Lt. Jos. took some more star shots.

13 December, 1942 (Sunday)

When the star shots were figured out, it showed us to be at 58° 12″—close to the town of Hebron. Worked on the put-put all day without success, so we tried to work the liaison set on the batteries but they were too weak. Then we pooled our covers and all slept together in the lean-to.

To most people familiar with the outdoors, the presence of wildlife for food and knowledge of a not-too-distant settlement would suggest a high likelihood of survival. But these were not outdoorsmen, and Northern Labrador was an inhospitable place for strangers, especially those from southern US military squadrons who lacked survival training in northern conditions.

> 15 December, 1942 (Tuesday)
>
> Didn't get up till noon because the wind was blowing. Our breakfast consisted of 9 cups of coffee and a package of Fig Newtons. The weather was very discouraging all day but late in the afternoon Nolan pulled the put-put through a couple of times and it caught. It didn't put out but 5 volts, however, but we were still very much encouraged. We ate a fairly big supper and went to bed early.

During World War II, under an arrangement with Great Britain, the United States Army Airborne Division patrolled the Newfoundland and Labrador coastlines with aircraft, and the US Navy with ships, to spot and fire on any Nazi submarine raiders. The military base at Goose Bay, covering 120 square miles of the southeastern portion of Labrador, was set up in 1942 and used as a takeoff or refueling point for American military planes bound for or returning from Europe. It was from this base that the crew of the downed Marauder were sure that help would come. In the meantime, they were coping with the severe climate, lack of food, and waning hope.

16 December (Wednesday)

Had to eat a cold breakfast (Chocolate bars, caramels) because the wind blew too much snow in our fire. Lt. Jans. and Jos. went fishing without any luck. Nolan changed the voltage regulators and got 26 volts long enough for me to pick up a couple of stations on the liaison receiver. The put-put stopped suddenly but we hope we know what is wrong with it so we may be able to get a message out tomorrow.

17 December, 1942 (Thursday)

The put-put put out, but we couldn't get the voltage to the front—must be a short somewhere. We tried to send on the command on Hq. batteries, but they were too dead.

The days were so windy that the crew just stayed in bed. The cold and lack of food were beginning to wear them down—*when* would help come?

21 December, 1942 (Monday)

Everything was really snowed in so we spent the day eating, thawing out blankets and planning the trip South. Lt. Janssen, Lt. Josephson and Sgt. Nolan plan to head South in the boat the first clear day.

Sending three men off in the seven-person dinghy seemed their only option. But could they overcome gale-force winds, the cold, short daylight hours, and high cliffs lining the volatile coastlines to paddle to Goose Bay and send rescuers back before it was too late?

22 December, 1942 (Tuesday)

Had a perfect day, the first clear day in over a week. We worked on the boat and cleared snow away from the lean-to all day. We ate a pretty big meal (a can of Spam and 2 κ Rations) with the 3 boatmen eating a little extra.

23 December, 1942 (Wednesday)

Got up at 7:15, got the boat ready and started carrying it. The wind was pretty strong and the boat was heavy, so we had a pretty hard time of it. We didn't get to the water until noon and then it took quite a while to find a launching place. We never did find a good one, but the boat was finally launched at about 1:30. The wind was pretty strong and tended to push them off-shore, but they appeared to be making some headway to the south the last time we saw them.

We had a hard time of it coming back against the blowing snow. We ate some peanuts and a caramel and went to bed.

Christmas Eve marked two weeks since their fateful landing. The day was windy and wet, and a fire in the aircraft caused Galm to burn his hand. It was difficult to remain optimistic, but survival depended on positive thoughts and actions.

25 December, 1942 (Friday)

What a Christmas! Mangini's feet pained him so much we had to get up at 3:00. He was in agony before that, but was better afterward, although his arches still pain him pretty badly. Got up again at 9:00. Galm went exploring, I massaged Mangini's feet and Weyrauch started fixing up the floor, which was in pretty bad condition from the fire . . .

I'm the only one who doesn't have anything wrong. Galm has a swollen hand with two blistered fingers, Mangini has two sprained arches and Weyrauch has a few blisters on his hand.

We are about to eat our "Christmas Dinner" and then go to bed.

The spirit of Christmas helped to lift their spirits. "G" Hodge, whose father was pastor of the First Baptist Church in Biloxi, Mississippi, made a special effort to count his blessings. Surely air patrols would be winging above them soon, searching for this lost crew.

27 December, 1942 (Sunday)
Started today as usual by treating the casualties. Mangini's feet are better, but we found a big blister on each foot. Galm is losing all the skin on one finger, including the nail.

Galm and Mangini spent the day drying blankets. Weyrauch finished cleaning out the back of the ship and I climbed the mt. to see if I could see anything out to sea. I also took a roll of film.

The weather has been perfect with the barometer reading 30.45 for more than 24 hours but it started going down this P.M and the wind is trying to pick up. I hope the weather stays good, tho, because we still have plenty of work to do.

The enforced diet is beginning to tell on us, but we will eat a little more tomorrow. Weyrauch spilled some coffee in the fire, it exploded and burnt the underside of the aileron off.

Their extended stay on the snow-covered tundra was wreaking havoc not only on the men's bodies, but also on one man's thoughts. A diary entry written by an unknown author between Hodge's entries on December 27 and 30 paraphrases a passionate passage from Welsh writer Richard Llewellyn's famous novel, *How Green Was My Valley*.[1] Perhaps the images helped one of the crew to recall greener pastures and fondly remembered pleasures. Following the passage, Lieutenant Hodge continued to record their condition.

30 December, 1942 (Wednesday)

Today was overcast with snow showers. Spent most of the day working on the invalids. Galm lost one fingernail and may lose another. I'm just thankful that his hand doesn't pain him.

Worked a little on the put-put and made some progress, but it was too dark to work much.

Got up a game of "500" Rummy which everyone seemed to enjoy. Had a delicious meal of a can of mushy apple juice and a can of tomato juice, of which we made a good soup by adding cracker crumbs. The boys have been gone a week today—God grant that they are still going strong.

New Year's Eve, 1942, was spent under overcast skies and light snow. The main problem now was fuel. They went to bed early to conserve the precious heat, each man left to his thoughts of what the new year might bring.

On New Year's Day, the wind and snow continued, but the rounded shape of the airship helped to keep it clear of snow and remain more visible from the air. Their heat consisted of a fire in a peanut can. Finally, Weyrauch managed, with the aid of a gallon of alcohol and glycerin, to remove the prop and receiver tank. Hodge dug out the oil drain. Fuel! Hot coffee, lemonade, bouillon! Dessert was the last can of date-nut roll, pieced out until noonday. The three boatmen had been gone ten days—the time anticipated for them to make the trip out.

3 January, 1943 (Sunday)

There wasn't much wind last night so we thought we would have a good day, but the wind picked up and it snowed all day.

The ship has a sheet of ice on it and is covered with snow. Besides that the drifts around the ship are higher and closer than they have ever been before.

We hooked up the hand fuel transfer pump and I am positive we pumped some gas over to this side but we couldn't get it out so we had to use the alcohol tonight to cook with. I got in too big a hurry once and caused a fire in which I burned my hand but not too badly—now we are all wearing bandages.

Found two Bouillon Powders in the Radio Operator's desk.

Spent a lot of time putting snow under our bed—there was quite a hole there, so we ought to sleep better tonite. It must be raining outside now, because this place is leaking all over and we only have a small fire, so it couldn't be melting ice on the wing. We keep praying for clear weather in the hopes that the boys got through—also to try out a new theory as to the location of the town of Hebron.

There were few rations left, and the men's strength was waning. Coffee was the only reprieve, bringing warmth and some feeling of fullness when there was nothing else.

7 January, 1943 (Thursday)
We've been here 4 weeks today. The entrance was blocked up this morning. As I was going in the ship I heard a noise and saw a little bird. We caught him, boiled him for a couple of hours and then made a stew by adding a Bouillon Powder—it was really delicious.

Galm started to go looking for Hebron, but the snow was too soft.

Mangini got outside for the first time in 13 days.

163

If we can't find a town or get the put-put going in 3 days, all we can do is wait for the weather to clear and pray that the boys got through, because we will be too low on food to do anything else.

God help us to find some way of getting out of here safely.

8 January, 1943 (Friday)

Today was the most strenuous for me since we got here. I tried to get to Hebron and still think I know where it is, but there are two mountains in the way.

I can feel myself growing weaker and we have less to eat every day. I don't know what we would do if we hadn't had the 3 pounds of coffee. We sit around and drink that and talk about all kinds of food, but I think we all crave chocolate candy more than anything else.

The boys dug out the back of the ship so if tomorrow is clear we will have one last try with the put-put and radio.

9 January, 1943 (Saturday)

Well, we got the put-put back in its place but it jammed again—so that leaves us with one possibility—that the boys get through. We had a perfect day here today, but are hoping that the boys are at Goose and the weather was [not] bad there. We don't have more food than for (one) good meal, but we are going to stretch it for six days. The boys here were down in the dumps today when the second of our three chances fizzled out in two days, but they seem to have snapped out of it pretty well. Also dried the blankets and remade the bed.

On Sunday they held a church service, as much to keep to tradition as to pray that rescue would come before it was too late. This day marked one month—thirty-one days—at Saglek Bay. Sunday's supper consisted of a slice of pineapple and two spoons of juice. "If we don't live to eat all the things we talk about," "G" writes, "we will at least have mentally eaten some of the best meals in the world."

The following days brought more problems. The oil gave out completely from one side of the aircraft, and what remained froze in the bitter cold. The short rations were affecting them mentally and physically, although they kept up their hopes through further church services and good debates—in one case arguing the respective merits of different candy bars.

14 January, 1943 (Thursday)
Clear day but with wind. We cleaned off the plane and waited, but nothing happened.

Late this afternoon we were playing cards when Weyrauch poured the gas too fast and caused an explosion which burned both his and my face, hair and hands. Our ration was four chocolates, but we're still making out pretty well.

After our little devotional we went to bed.

The diary entries become shorter and assume a tone of quiet desperation. Still, Lieutenant Hodge manages to keep the record impersonal.

18 January, 1943 (Monday)
Cold and clear—my watch stopped, so we didn't get up til noon. Must be a little warmer, because we got a little oil. Today was our first completely foodless day, but spirits are still pretty good. It's surprising the amount of punishment the body and mind can take when necessary. We are still in pretty good physical condition tho rather weak, and are still capable of talking cheerfully of food and home, tho there can't be too much hope left.

They let the oil run all that night to get enough heat to make their "hot water special coffee" for breakfast, all that was left. Again, the men stayed in bed until after noon. Another small fire broke out, damaging a cushion. The snow continued to fall and the wind blew it up against the door. The men now resembled hibernating bears—gaunt, hungry, and nearly catatonic.

23 January, 1943 (Saturday)
Spent a miserable night—everybody was crowded and nobody could get comfortable. Had a good day but everybody is getting pretty discouraged, altho the conversation was pretty good. We haven't really felt famished but we really are weak. It gets me to see the boys start to do something and then stop for lack of will power to go on. Weyrauch has developed a case of piles which is bothering him a good bit.

26 January, 1943 (Tuesday)
Overcast but fairly calm. Each day we don't see how we can last another day, but each time we manage to go on. We all smoked a pipe of tobacco this morning and Galm got really sick, and I felt pretty bad, but we came out of it pretty well.

3 February, 1943 (Wednesday)
Spent a solid week in bed. Today Weyrauch died after being mentally unbalanced for several days. We are all pretty weak, but should be able to last several more days at least.

The journal entries end here.

DETECTION

On April 9, 1943, two Inuit hunters traveling by dog team across Cape Saglek, thirty miles north of the settlement of Hebron, discovered the wreckage of the Marauder. They peered under the rough lean-to made from the engine and wing covers draped over the starboard wing and saw three frozen bodies. A fourth was found later outside in the snow.

The wreckage was reported to E. B. Gillingham, a Newfoundland Ranger stationed at Hebron.[2] He traveled to the site and filed this report of his discovery.

Newfoundland Ranger E.B. Gillingham, Hebron Detachment, #107, report to the Chief Ranger, St. John's Nfld., dated Monday, April 12, 1943:

On Monday I went to Tikerargut to go to the scene of the aeroplane crash. I left there early in the morning and arrived at Saglek Bay at 12 noon. We motored up the bay for an hour and then went ashore on a neck of land.

After walking for approximately one mile we came to the scene of the crash. The plane is upon level ground although there is a lot of rocks about. On one side of the plane at one mile and a half distance is a mountain called the DOME. This is a mountain of 2,000 ft. in height. From this mountain Hebron is visible.

The plane is lying with her nose due North-east although when first she struck she was going due North. This is proven by the wreckage found. The cause of her turning after covering some 200 yds., was caused by her propeller having jammed in a rock. The piece is still in the same position as when it jammed. The numbers on the plane are as follows: US Army

Saglek Bay. The object in the centre is the base of the old Pinetree DEW Line radar site, closed in 1976. (Courtesy Clarence Simonsen)

B-26-B2, American Air Force, Serial No. 41-17862, crew weight 1,000 lbs., crew consisted of seven men.

Around the plane is wreckage thrown for a good distance and underneath the body is torn for 14 feet. Near at hand is a tin helmet set into an ammunition case. This apparatus was used for a stove, the helmet was used as a pot. Under one of the wings a tent was constructed from parts of parachutes and engine covering. There were four bodies found in the tent although the crew consisted of seven men. The condition of the four men that were found seems to point to starvation rather than being frozen. One of the men had a diary and in it was stated that they were warm but were so weak from hunger that they could not hunt. They were found huddled together in the tent.

As far as can be found three of the men left in a rubber boat to come down the coast (This also was taken from the diary). At this season of the year the weather is cold and stormy so that they most certainly must have drowned.

The plane crashed about the 10th of December, and the men lived till the first of February.

The man who found the wreckage was Joshua Kaujasiak. He found the wreckage when returning from hunting. He came to Hebron and notified Mr. Simpson accordingly.

Certified correct, B. Gillingham, Reg't No. 107, I/C Hebron Detachment.

On receiving Gillingham's report, US Army personnel took off from Presque Isle on April 15 in an AT-7 Beechcraft Navigator, flying via Goose Bay and Crystal 1 (Fort Chimo, Quebec, now called Kuujjuaq). There they boarded a Noorduyn Norseman for the last two hundred miles of the trip, landing on the harbor ice at Hebron on April 17. The next day, US Army Air Corps Major Norman D. Vaughan and Lieutenants Norton, Holmes, and Burkhalter left Hebron with Inuit men and dog teams, traveling overland and along shore ice to arrive on April 18 at Cape Saglek.

As reported by Gillingham, three bodies lay in sleeping bags inside the lean-to beneath the right wing. The fourth was outside. These men were identified as 1st Lieutenant Grover G. C. Hodge Jr., Corporal James J. Mangini, Corporal Frank J. Galm, and Sergeant Russell Weyrauch.

Lieutenant Norton and the Inuit men returned to Hebron the next day with the four bodies, leaving Major Vaughan, assigned as the officer in charge of special mission to plane #1786, with the downed aircraft "to salvage certain valuable equipment" and transport it to Hebron by dog team. According to Major Vaughan's report, submitted April 29, 1943, personal effects also included two envelopes belonging to crewmen Janssen and Josephson, "but their bodies were not found at the wrecked aircraft."

Vaughan's follow-up report of May 3, 1943, states the names, ranks, and serial numbers of the bodies found with the aircraft. He adds, "Found diary of Lt. Hodge, Captain of airplane, which states that on December 10, 1942 the plane took off from BW #1 for Goose Bay, Labrador. The plane got lost, ran out of gasoline, and made a crash landing. No one was injured. The only obvious damage was the torn and sprung bomb bay section and eight broken and bent propeller blades. Copy of the diary entries from December 10, 1942, to February 3, 1943, are herewith attached." From Hodge's diary it was learned that Janssen, Josephson, and Nolan had left in the rubber dinghy on December 23, 1942, to search for Goose Bay. Major Vaughan concludes, "No word has been received from these men. This accounts for all seven men on the plane."

Lieutenant Burkhalter arranged funeral services with military honors at Crystal 1 on April 23, 1943. Chaplains of various faiths officiated, and the four bodies were interred in the US Army cemetery plot. Major Vaughan, Lieutenant Holmes, and Lieutenant Norton returned that same day to Presque Isle via Goose Bay, with their part in the mission accomplished.

THE AFTERMATH

Report followed report on the discovery. On May 4, a memo was sent by Major Vaughan to the commanding general, North Atlantic Wing, ATC, with receipts for the equipment brought back to Presque Isle. Although the accident had occurred on Canadian soil, the aircraft was being operated at the direction of the USAAF. Salvage and reclamation of the aircraft were therefore turned over to the subdepot of the air service command at Presque Isle Army Air Field.

Major Vaughan also noted that, on the forward side of the left panel in the bomb bay of the B-26, some screws, wires, and equipment supports were disarranged "as though equipment had been removed." This is where the Identification Friend or Foe (IFF) radio instruments were usually installed (on wartime bombers going into combat, these instruments were installed so when they returned to base after a bombing raid they would be confirmed as friendly and given landing assistance), but no sign of this equipment could be found. (However, "G" Hodge's brother-in-law, John Rose, states, "We have yet to confirm that the aircraft was so equipped, since the units were in short supply and may well have been omitted from an aircraft not yet operating in a 'hot' combat zone.")

On May 31, Major Vaughan (signing as former summary court officer) listed the personal effects of the Marauder crew members, stating that the materials were being turned over to the current summary court officer for further disposition.[3]

The fate of the other three B-26s returning from Greenland the night of December 10, 1942, was recorded by Goose Bay control tower officer Captain Robert E. Griffin. He had made contact with the first one and successfully guided it in. The second was believed to have crashed into the Atlantic. The third went into a dive when it encountered a storm and flew back over the water to BW-1 in Greenland. Now, the question remained, what had happened to the three boatmen? Although it was generally supposed that they had drowned, an intriguing report surfaced that begged investigation.

The headquarters of the RAF in Canada during the war was in Dorval, Quebec, operated by No. 45 Atlantic Transport Group. All correspondence related to wartime anomalies on Canadian soil went through their offices. In a report to their commanding air officer dated April 30, 1943, RCMP Superintendent K. Duncan wrote that on April 27, Native trappers had spotted the tracks of two men on snowshoes and one on skis approximately seventy miles east of Fort Mackenzie: "Indian was given facilities by Americans to return to position and endeavor to locate and trail these tracks. US Army have no personnel lost in that area and require information re: possibility Canadian air crew being lost in vicinity."

On May 3, Superintendent Duncan indicated that, if necessary, a plane would be sent to search the area. On May 19, Duncan informed Flight Lieutenant Sherwood, Air Operations at Presque Isle, that "these men were believed to be survivors of a plane crash in that area," and wished to ascertain whether any planes had been reported missing in that vicinity. The downed Marauder was then brought to his attention.

Fort MacKenzie lies inland on the Quebec side of Labrador. If by chance the boatmen had sailed a couple of hundred miles south and then gone inland, they might have missed the Hudson's Bay post of Fort Mackenzie by just a few miles. It is possible that they were outdoorsmen, while the four who stayed with the aircraft were not, and they'd managed to survive on their rations or shot small game such as rabbits.

Intriguing reports continued to come in, and Duncan submitted this information:

> Some Indian women reported that they'd heard the sound of a high-powered rifle a few days after the original tracks were found. [However] as the area in question is heavily timbered with small lakes further search by Indians is not practicable.

Normally, the only portable firearms on board the B-26 aircraft would be the .45 caliber officers' handguns and a .22 caliber survival carbine. It is possible, however, that someone could have brought along a personal firearm, as no checks at customs or otherwise were made in this regard. Perhaps someday the bones of the three boatmen might be found, alongside their American firearms.

FAMILY MATTERS

When Hodge's family (sister Alma and her husband John Rose, and brother Lieutenant (Ret'd) William T. "Bill" Hodge and his wife June) attempted to retrieve the diary, as well as the camera and film noted in the December 27, 1942, diary entry, these items were not to be found. The diary reference was the only indication of the existence of a camera and film. Major Vaughan's list of items and Inspector Gillingham's report both indicated Hodge's diary with its dated entries.

On September 14, 1945, Lieutenant Bill Hodge, then assistant operations officer at Hickam Field, Honolulu, Hawaii, chanced to meet a pilot who had been on the mission to take the bodies from Hebron to Crystal 1 for burial. Bill made a written record of their conversation: "Had they gone about eight miles inland they would have found a trail leading to the settlement," the officer had said. "Or, had they started down the beach, within 24 hours they'd have met up with Natives. Had they climbed to the top of the highest hill on a clear day, they might have seen some huts. *But* they had been instructed to stay with the plane in the event of a crash, and that is what they did in the hope of a military rescue."

The day following the crash, when the crew had taken sun-shots and discovered they were only thirty miles from Hebron, they must have thought it was just a spot on the map. Actually, it was a Moravian mission, occupied throughout the year (until 1962) by a German missionary. The village was also inhabited by the Scottish Hudson's Bay post manager and two employees, the Newfoundland Ranger, a white trader and his wife, and some Aboriginal residents.

Photographs in Clarence Simonsen's collection, taken by Calgary pilot Roy Staniland, show nearby hills, referred to as Dome "A" and "B," from which the settlement could easily be spotted. In fact, Hodge's diary entry of December 13 states they knew their location in relation to Hebron: "It showed us to be 58 degrees, 12", close to the town of Hebron." Because of the proximity of the town, the officer was critical of the Marauder crew for making no attempt to hike there. The trail, however, was rough, boulder-strewn, and necessitated climbing over several two-thousand-foot hills— all in the worst weather by men who were neither dressed for the cold nor carrying proper food or equipment.

Adding insult to injury, the officer continued to reprimand the actions of the men, now being called "The Lost Crew," to Hodge's grieving brother. "A few days after landing Hodge wrote they'd seen about 50 seal and a fox and figured they would have plenty of food if they needed it. They did not know that the animals were migrating southward and so made no effort to utilize this available food supply. The three men setting out in the dinghy—contrary to the 'stay with the plane' theory—were foolhardy as it was certain that the boat would become punctured by ice." Also, U-boats were wreaking havoc in Atlantic waters.

"The pilot I met in Hawaii was incorrect in some of what he told me," Bill Hodge says, "but how does one deal with statements that were made, but are wrong in part or whole?" Yet Bill Hodge later verified that the village *could* be seen from the mountains where the radar station is later located. The buildings are still at Hebron. He tried to see them on a crystal clear day and could make out an irregularity, but argues there is no way it would have been recognized as a habitation unless one was told and knew where to look. "They did make an attempt to hike to it," Bill Hodge affirms. "The diary tells of my brother trying and giving up in waist-deep snow. The trail leading to the settlement would have been no more than a path through the rocks and impossible to see under snow. As for the 'beach,' there is no beach as such. There are stretches of sand but more often the water's edge is rocky or sheer cliffs down to the water. After the shore ice has hardened thick enough to bear a person's weight, the Inuit walk on it. We can't fault untrained, very weak men, for not trying that."

Further, the pilot Bill Hodge met in Hawaii said that each man had letters to his family on his body, and that there were letters left by those who went on the boat. "I can only guess that they were destroyed in the cover-up that continues to this day," Bill Hodge writes, "and the manuscript diary might have been destroyed because the booklet contained 'G's letters. The dates of death of all are still given as December 10, 1942."

For three years, the bodies of the four airmen remained interred in their graves at Crystal 1 (Fort Chimo) until being removed to Fort Pepperell Cemetery in Newfoundland. When that was declared a "temporary" cemetery, graves of all US armed forces personnel were exhumed and shipped to the United States. On November 20, 1947, the remains of 1st Lieutenant Grover C. Hodge, Jr. were forwarded to Biloxi, Mississippi, where he was honored by a hometown service at his First Baptist Church. His grieving fiancée, Velma Fern Carter, had died the previous April and her body rests nearby.

The family received some of "G" Hodge's effects in the 1940s, but were unable to see a transcript of the diary, or get official acknowledgment that it even existed. Their first view of excerpts from the diary was in a story titled "Dead Man's Log," published by a popular magazine, *Cavalier for Men,* in January 1954, fully eleven years after the accident, which gave fictitious names to the crew members. The story was later published using the men's real names. The family's next view of the diary entries was in a *Miami Herald* article titled, "Diary Records Last Days of Doomed Crew," published on February 14, 1960.

By 1969, the Hodge family had still not been provided with the original or a copy of the diary, or even a photograph of a page in order to verify their brother's handwriting. In frustration, they appealed to their Florida congressman, Paul G. Rogers, for assistance. His efforts resulted in the "declassification" of a typewritten transcription.

An accompanying letter from Colonel Kenneth Dill, USAF Congressional Inquiry Division, Office of Legislative Liaison, Department of the Air Force, Washington DC, dated January 9, 1970, states that after extensive archival searches by officials in air force headquarters, the original diary could not be located. In fact, they were unable to establish what had happened to the original writing following its discovery and attachment to Vaughan's report filed in 1943. Colonel Dill also notes:

It was against regulations in World War II for servicemen to keep diaries and that of Lieutenant Hodge was apparently given a security classification after it was found. During the war either the original or a copy (we cannot determine which) was in the Intelligence and Security Division, North Atlantic Division, file 'Search and Rescue #1'. The present whereabouts of the file is unknown.

Biographical data on Major Norman D. Vaughan (herein called Lieutenant Colonel Vaughan) was attached, as he had supervised removal and burial of the bodies, as well as later disposition of the crew's personal effects. Vaughan's past exploits are "so tinged with color and excitement as to read like fiction from the pen of a latter-day Jack London—but more Spartan," Dill states, adding erroneously that Lt. Colonel Vaughan had died in 1960.

QUESTIONS AND ANSWERS

The poignant question of why the Lost Crew had waited for over two months for a rescue that never came, even though "G" reports the weather being clear for many days, was answered by Colonel Dill, referring to a report contained in a history of the North Atlantic Division, Air Transport Command:

> B-26B, Serial No. 41-17862, took off from BW-1 (Narsarsuaq) Greenland at 12:57 Greenwich Meridian Time on December 10, 1942, to return to Goose Bay and was reported overdue, along with two other B-26Bs (missing on December 8) "but proper search was prevented by weather." At that time no formal 'search and rescue' organizations were available for this air route.

Continued pleas from the family to the military for the retrieval of the original diary and other missing effects went unanswered. Finally, on June 3, 1993, Alma Rose wrote to Congressman Tom Lewis, Palm Beach Gardens, Florida. From his office they received a weather and accident report—the latter containing large holes where portions of the text had been removed with an Exacto knife.

Through Lewis's congressional efforts, Alma and John Rose also got a chance to speak by telephone to the very-much-alive Norman Vaughan at his home in Alaska in 1993. The conversation was brief. Vaughan was busy raising funds for his proposed Antarctic expedition to climb Mount Vaughan, named for him by Admiral Byrd in the 1920s, to celebrate his eighty-eighth birthday that year.[4]

When General Curtis LeMay (commander of the Strategic Air Command in the late 1940s and early 1950s, and later air force chief of staff) heard of the diary in 1948, he publicly lauded the men's courage and endurance, but expressed puzzlement over why they hadn't done more to help themselves. General LeMay subsequently applied the lesson learned from the Labrador Marauder tragedy to his SAC bomber crews, resolving to give his crews a fighting chance if forced down, regardless of the environment in which they found themselves. The USAF Survival School was activated in March 1996 as the 336th Combat Crew Training Group (Survival). Rigorous training exercises were developed for survival when flying over every type of global terrain—land, inland water, and sea.[5]

EARLY WARNING SITES

The lonely shores of Labrador became active after the war, with construction of early warning stations there and elsewhere across the Arctic, such as the Pine Tree Line for the Canada–United States Radar Extension Plan, the Mid-Canada Line, and the Distant Early Warning (DEW) Line. The Saglek

Pine Tree site, constructed in 1951 by the US Army Engineers, became operational in the fall of 1953. This necessitated building a five-thousand-foot asphalt runway, and so the wreckage of the B-26 Marauder, *Time's a' Wastin'*, was pushed aside.

New generations of military and civilian personnel who served at "The Sag" became familiar with the drama that had once taken place here, as they viewed the old bomber lying on its back, with metal wing pieces flapping in the wind and odd bits, not already taken by souvenir hunters, strewn about the rocky ground. The approach to this new runway challenged pilots with the dangerous downdrafts that swept in from the steep surrounding hills, proving that Lieutenant Hodge had landed his plane at the best possible area for many miles around.[6]

A plaque was erected near the remnants of the doomed aircraft by personnel of the 924th AC & W Squadron, and dedicated "to the ill-fated crew of the B-26 which crashed near this spot in December 1942—a tribute to their heroic effort to survive." The plaque is now housed inside the radar site building. The story of the final days of the Lost Crew is further recorded in the archives of the B-26 Marauder Historical Society at the University of Akron, Akron, Ohio (moved in 1998 to the Pima Air and Space Museum, Tucson, Arizona).

Larry Wilson, a Canadian historian, became interested in the story and helped the family to visit the Saglek site. His website shows photos and tells the story of past and present occupations of this site.[7] In August 1998, fifty-six years after the crash, the Hodge and Rose families came to Saglek Bay to dedicate the new memorial.

A TRAGEDY - VINTAGE 1942

PRESENTED BY

THE 1932nd COMMUNICATIONS SQUADRON (AFCS)

GOOSE BAY AIR BASE, LABRADOR, CANADA

En route, their aircraft stopped to refuel at Nain, now the northernmost village on the Labrador coast. There they spoke with Gary Baike, director of the Labrador Inuit Association Centre, who told them that his grandfather was one of the men who'd found the plane; his uncle had helped Major Vaughan bring out the bodies. His uncle had also informed Gary that the diary notes were written in the pilot's logbook. In the 1990s, the Hodge/Rose family personally talked to the USAF major who was assigned the follow-up task of trying to locate the logbook. "He reported, as did many others, that it was probably in a set of missing files from the North Atlantic ATC Operations area," writes John Rose.

In a letter to Canadian researcher Clarence Simonsen dated August 3, 1993, Dr. Alex Douglas, Director General History, Canada National Defence, presents this matter-of-fact view of the mystery of the missing diary:

> One reason why Alma Rose Hodge is having such difficulty in locating the original diary might be that it may no longer exist . . . From what you write, it appears to have survived the crash but it may have been lost or destroyed in the intervening half century. Or perhaps it is languishing, unknown, in an archival repository in Canada or the United States.

To date, no one in Lieutenant Hodge's family has had the opportunity to view and verify the handwriting or the contents. Only typewritten transcripts of entries nearly identical in length and style that describe daily living conditions (with American and Canadian versions differing somewhat) have been produced.[8]

THE VANISHING NOSE ART

Clarence Simonsen of Airdrie, Alberta, is an expert in the unique field of aircraft nose art, with a number of books and articles published on the subject. His interest in the Saglek saga came about through a reunion of the 319th Bombardment Group and its history, which includes the story of "The Lost Crew."

Simonsen obtained some pieces of wing-skin from the B-26 *Time's a' Wastin'*, made a painting of the doomed aircraft, and sent it to Lieutenant "G" Hodge's family in the USA, along with further sketches of the crash site. "The nose art, *Snuffy Smith – Time's a' Wastin'*, had been painted on the aircraft either in Omaha where they picked up the plane in September 1942, or at Fort Wayne, Indiana, their next assigned station, by an artist in the 320th Bomb Group, 440th Squadron," Clarence Simonsen affirms. "But the nose art just disappeared once the US Navy went there on their surveillance in the 1950s, and that's another one of the puzzles. The result of souvenir hunters? Or attempts to disguise/destroy the aircraft?"

William (Bill) C. Parrott, a military aviation researcher and Director of Crown Lands, Government of Newfoundland and Labrador Newfoundland, suggests that the artwork and other aircraft parts could have disappeared "through a combination of human salvage and scavenging for scrap and souvenirs, along with the high winds which would have caused the remaining wreckage over the years to tumble about and break up to the point where it could be literally carried by the wind."

In 1952 the United States Navy went to Saglek Bay, where the intact Marauder rested with only a rip in its belly and a damaged prop that had hit a rock, and apparently destroyed it. Bill Parrott is not surprised at reports of the aircraft being burned. "Burning of aircraft that were not to be removed was a common practice of the USAAC in Newfoundland and Labrador during the war, and [was] probably the eventual fate of the B-26 after the salvage was completed."

THE QUESTIONS CONTINUE

The search for the vanishing nose art has led Simonsen to ask more questions, particularly about the activities of the United States Navy following their visit to Saglek Bay in 1952. "The US Navy then ventured north that same year and surveyed Home Island just across the bay from Cape Chidley, but there are no official records of them finding the Nazi weather station installed there on Canadian soil during the war." He asks, "Isn't it odd that the US Navy hydro-graphic team in 1952 visited both of these two dots on the Labrador map, Saglek Bay and Home Island? The Nazi weather station is centered right in the middle of a 'wagon wheel' of the American flight routes. That wasn't a coincidence."

It is known that a weather station was established by enemy forces in October 1943 on the shores of Labrador at Cape Chidley; as well, there were other radio-operated stations set up throughout the North Atlantic region around the time of the B-26 crash at Saglek Bay. American aircraft often received misleading information. The diary entry of December 10 stating, "Lt. Josephson gave me a heading to get back on course, but we know now that it was too big a correction," seems to place Josephson at fault. But had the Marauder been misled by the Germans, causing the crew to fly north to their doom rather than south to Goose Bay?

Dr. Alex Douglas disputes this idea but leaves a possibility open for consideration:

> We have no way of knowing what radio signal the crew of the USAAF B-26 would have picked up.
>
>   It could have been another weather station but it does not seem likely that it was a German one, as the evidence indicates this [at Cape Chidley] was the only station they set up in Labrador . . . We do know, however, that aircrew with the British and American wartime ferry services delivering aircraft via the

North Atlantic route sometimes reported receiving signals of mysterious origin, apparently designed to direct them off course.

"I'm not trying to create a mystery," Simonsen concludes. "I'm trying to ask why are the Rose and Hodge families still having problems getting information. No one wants to talk about it. What is the mystery here that's causing the Americans to still keep their mouths shut?"

It is still unclear why Hodge and his crew were left to suffer a slow death marooned near the savage shores of Saglek Bay. It is known that the early American military flyers had little aviation experience or survival training. They were simply told, "If you lose an engine, stay in the plane and go straight into the ocean, because if you bail out you'll live maybe two minutes at the most in the ice cold water of the Atlantic." Many did just that. They were young, healthy, energetic, patriotic, and eager to fly to combat. Nothing would happen to *them*. The nose art on their aircraft reflected this type of thinking—America's involved, "time's a' wastin."

Were they left to their fate because nobody knew they were there? Or because, as noted in Colonel Kenneth Dill's report, "at that time no formal 'search and rescue' organizations were available for this air route"? Did the eager young fliers know this? One supposes not, for Hodge and his crew never gave up the hope of being rescued by their country's forces.

Although many aircraft went missing in action during wartime, Bill Hodge still feels bitter that reconnaissance planes were not sent along the coast. John and Alma Rose agree. "Try looking into why the allies were so ill-prepared to enter the war," they write. "Investigate why we Americans so often substituted youthful exuberance and enthusiasm for solid training and experience for our aircrews . . . and how dearly we paid for these substitutions! Also look into why 'G' and his crew members were never paid for a single day of military service past the December 10, 1942, crash date. Was it because of the diary's being classified that Army Finance was never told of the crew survival past the crash date—because they 'did not have a need to know'? Find out why so many crews were abandoned after they were forced down and died of their injuries or starved to death, as did 'G' and his crew."[9]

"Everyone has interest in this case," Clarence Simonsen concludes, "but no one will talk. No one will help to provide answers. So then you end up saying, 'It's just a plane crash.' Yes, but then why all this intrigue?"

UNDERWATER PHANTOMS: SECRETS OF THE DEEP

Aviation in itself is not inherently dangerous.
But to an even greater degree than the sea,
it is terribly unforgiving of any carelessness, incapacity, or neglect.
—author unknown

Can phantoms rise from the dead? Stories persist of ghostly crews returning to communicate with earthly mates, or lost planes being resurrected from the deep to once again embrace the sky.

While salvage of airplanes found long abandoned underwater is fraught with jurisdictional problems and involves huge amounts of money, interest in the recovery and restoration of these ghostly galleons of the air is unending. Salvagers worldwide who love these old aircraft will fight through multitudinous rules and regulations to gain permission for their recovery.

The US Navy is the most sticky of all military sectors in this regard, enforcing policies on sunken navy vessels and naval aircraft wreck sites in order to protect its historic properties. The US Air Force is more lenient regarding salvage of wrecks; however, any human remains discovered at wreck sites must be removed before salvagers can proceed.

The Canadian government is more accommodating toward private sal-vagers who apply to retrieve sunken military aircraft. According to Major Dan Bellini, director of disposal, salvage, artifacts, and loans at Canada's Department of National Defence headquarters, while the Crown maintains ownership of all Canadian military wrecks, the department does grant sal-vage rights and title to individuals.[1]

These rules—often hotly contested—have led salvagers to become cau-tious of sinking money into sunken aircraft and ships before the paperwork is complete. But there are cases where the images of those who have per-ished in downed aircraft return, notwithstanding the government's rules and regulations about ownership and hands-off attitudes, to visit their sister ships and former crew mates. Some such tales even graduate over time from folk-lore to reality.

These stories involve airmen and airships that met their end in watery graves, but were not destined to remain there.

THE FATEFUL FLIGHT 401

At 9:00 PM on December 29, 1972, an Eastern Air Lines, Inc. TriStar "Whisperliner" L-1011 (Plane #310) left JFK airport, New York, on Flight 401 bound for Miami. Following a rare and bizarre sequence of events, the big airliner crashed into a Florida swamp, killing all thirteen crew members and eighty-eight of the 163 passengers.

Soon after, Captain Bob Loft and Flight Engineer/2nd Officer Don Repo appeared as ghostly images to a number of crew members and pas-sengers on other Eastern Air Lines TriStar flights. These sightings were reluc-tantly reported by airline employees, who feared being ridiculed, sent to visit the company's psychiatrist, or perhaps even fired. By the time these

above: The Eastern Airlines Flight 401 crash site. (Courtesy Don McClure)

left: Don McClure, a member of the accident investigation team at the Flight 401 crash site in the Florida Everglades. (Courtesy Don McClure)

accounts were published in the us *Flight Safety Foundation Newsletter* in 1974, the stories had become significant and believable.

Follow-up phenomena soon convinced the public, and even many airline executives, that the ghosts of Plane #310 were not at rest. When non-structural components such as radios, electronic and avionics instruments, and galley equipment ranging from elevators to stainless steel ovens were salvaged from the wreck and used on other Eastern Air Lines (EAL) aircraft, the ghosts followed. These parts were costly and could be transferred to other aircraft, even though it was traditionally considered unwise—and was often forbidden—to salvage parts from a sea- or airship that had been involved in a disaster. (According to a mariners' superstition, metal should never be salvaged from a ship where there has been sudden, catastrophic loss of life, on the theory that the spirits of the dying could be "trapped" in the electron field of the metal.)

Various people in the aviation industry corroborate, however, that such parts *were* frequently salvaged from wrecks. "This was in the 1980s—maybe it isn't done now, but it was then," one aviation mechanic wrote to an Internet website where the crash of Flight 401 is still avidly discussed. "We were told to retrieve any parts without damage as they were incredibly expensive and would be used on another aircraft. This was a common practice when a plane crashed."

One day, two EAL mechanics were working in the lower galley of a L-1011, filling fuel tanks. The power to the aircraft had been completely shut off as a regular safety procedure, when suddenly the galley fan came on. The mechanics vamoosed. Another mechanic reaching around for his screwdriver had it placed neatly into his hand—by no one, for he was alone. A mechanic was advised of a trouble spot in the hydraulic system. As he turned to thank his informant, the "person" vanished in front of his eyes. Maintenance crews noted that sometimes the light in the hydraulics service area would not stay on, and mechanics refused to work on this particular aircraft after dark. The rumor spread that this aircraft had, indeed, received some parts from the doomed L-1011, Plane #310.

The stock department, where serialized parts were catalogued along with their origins and destinations, was finally told to trace all parts that had come from Plane #310. They would subsequently be removed from whatever aircraft they were now on, even if working perfectly, despite the cost. The spirits had to go. The appearances were too frequent, too believable, and had occurred in front of too many people to be dismissed.

Canadian technicians came to know of these phenomena as well. On an Internet discussion page regarding Flight 401 phenomena, a person advised that in 1989 he had attended the Metropolitan Toronto Reference Library to research the technical parts manual for the Lockheed L-1011:

> I thumbed through the enormous volume (about three Toronto phone books thick) until I reached an index of salvageable airplane parts that had been removed from the doomed jet, Flight 401. It itemized each part and an asterisk outlined a footnote that included a summary of John G. Fuller's book, The Ghost of Flight 401. It explained the association between the airline parts and the subsequent hauntings that took place amongst the Eastern fleet. I feel that if the Lockheed Corporation acknowledged the story in their TriStar jet parts manual that it must be a true story. [1]

Understandably, this letter prompted responses from dissenters who had serious doubts that a technical manual would contain a notation about a ghost story, and that a public library would stock something as obscure as a parts catalogue for a particular type of aircraft. (At the time of publication, checks with this library indicated twelve books under Lockheed in their collection, but no such manual.)

But a former EAL mechanic definitely recalls parts being "secretly removed and installed on Air Canada's fleet of L-1011s out of Kennedy, and subsequently Air Canada's parts were installed on Eastern's fleet." Following those maneuvers, he says, no further incidents were reported.[2]

The "ghost" plane was totally scrapped in the early 1990s, and the majority of L-1011s are out of service today, as is Eastern Air Lines. When the company went bankrupt, most of EAL's aircraft went to Marana, Arizona, along with their parts stock, to be stored in a hangar until dispersed to new buyers. Delta Airlines purchased the remaining L-1011 fleet in 1991 from EAL's liquidation, as well as most of their parts. As recently as 2001, American TransAir was still running about twenty former Delta and TWA L-1011s. One of their flight attendants was told that at least one aircraft still contained parts from the "Flight 401" plane, and a few attendants and pilots reported seeing ghostly faces reflected in the lower gallery and on the instrument glass in the cockpit.

Parapsychologists coin these as "crisis apparitions," which seem very real until they suddenly vanish. The spirits are considered to be in trauma, trapped in the former world because of their "sudden death out of normal time" and blocked from moving along on their spiritual path. While Captain Loft's image was seen by fewer people, and then not at all after a year's time, Flight Engineer Don Repo's ghost continued to appear. It was decided that he needed a "soul rescue," which was done in New York through séances.[3] His image was not seen again, although it returns in one's imagination whenever good ghost stories are told.

THE PHANTOM LOCKHEED HUDSON

Canada also has its share of stories involving airmen and airships that met their demise in watery graves, but were not destined to remain there. In January 2002, the *National Post* reported a story that had long excited Greenwood Military Aviation Museum personnel. Down in a two-and-one-half-mile-long, forty-foot-deep gully in the brown bog-fed waters of Gaspereau Lake, Nova Scotia, lay an object with a shape and size resembling a large aircraft. Curious and cautious scuba divers pushed their way down into the murk in attempts to check out the phantom, said to be a Lockheed Hudson.

To find this World War II twin-engined bomber, intact and awaiting discovery, would be rare and exciting. Just five such restored models exist today, and only one in Canada, on display at the North Atlantic Museum in Gander, Newfoundland. The tapered midwing monoplane has an all-metal stressed-skin construction and a transparent nose for bomb-aiming. If this aircraft in Gaspereau Lake was in good shape, it could be worth as much as five million dollars when restored to airworthy condition, although restoration would be costly and time-consuming.[1]

And so began the intensive search, permitted by the Nova Scotia government and overseen by the Greenwood Military Aviation Museum, which holds the salvage rights. Retired Major General Ian Patrick, chairman of the board of directors, is emphatic about the importance of this discovery and avidly supports the museum's efforts to retrieve the silt-covered "bogey." But, as Patrick reports, "this aircraft does not want to be found."

In March 1943, says local legend, a Lockheed Hudson returning from a U-boat hunt near Shelburne, off the south coast of Nova Scotia, experienced engine trouble and came to rest on the lake ice, fifty miles northwest of Halifax. This was corroborated by a World War II radio operator who'd noted in his logbook that a Hudson had force-landed on a Nova Scotia lake. There was no loss of life to the four-person crew.

Repairs were made to the engine, but during the test run the vibrations caused the ice to crack and the aircraft began to sink. Before it was lost, a technician was ordered to remove the autopilot and important gear. He had just begun when the water started to rise. The door stuck, and he barely made it to safety through the overhead hatch before the aircraft sank. With no known fuselage number, positive identification of the submerged aircraft became impossible.

The Lockheed Hudson Mark V. (Courtesy Major Gen. (Ret'd.) Ian Patrick)

Gaspereau Lake, once a natural body of water, had been dammed in 1942 for a hydro project and remained flooded. When the lake levels receded in 1998, military search and rescue technicians spotted the image of the aircraft in the dark and dirty water. The location was recorded by Sergeant Jean Roy and Master Corporal Darrell Cronin, who described seeing two vertical stabilizers on the tail and a gun turret on the upper fuselage, which fit the profile of a Lockheed Hudson. Before they could record their sighting on a map, Roy, Cronin, and four other crew members were killed in the crash of their Labrador helicopter in Quebec's Gaspé Peninsula. The coordinates they'd recorded also vanished, and no other air crews could find a submerged aircraft that fit that description and location.

In 2000, underwater video footage was taken of an aircraft identified as either a Hudson or a Ventura, and the longitude and latitude recorded with GPS instruments. But then the photographer left town, and two years passed before he returned to verify his images. Major General Patrick states that a number of dives were undertaken, but all met with problems such as "unusually bad weather, unseasonable high water, inexplicable broken equipment, and motors that would not start." Patrick says, "The curse of the Hudson was now a reality. It did not want to be found."

In 2002–03, the Greenwood museum called Satlantic, a Nova Scotia company that expressed an interest in testing some new technical equipment in realistic circumstances. This included a laser underwater camera imaging enhancer (LUCIE) to penetrate through the murk to locate and identify submerged objects, whose images are then relayed to a surface video screen. "But, the aircraft eluded us once again," states Patrick. "New equipment developed early problems, the shallow water took its toll of boat propellers, and the discipline of covering every area thoroughly was lost in concentration on most probable areas."

Army divers arrived, planning to use the search as realistic training. "These divers also suffered the fate of those who had previously searched for this aircraft," Patrick says, "such as near collisions with large rocks in the murky depths; almost being trapped when unstable boulders, disturbed by the divers, fell among them; having wet-suits flooded, allowing ice cold water ingress next to the skin; blocked ears in experienced divers; broken valves on air tanks; broken propellers and motor masts; blown bearings on boat trailers; and a host of other events that slowed or prevented searches by these professional underwater experts."

In the autumn of 2003, water levels dropped to an all-time low and it was hoped that the large aircraft could be found. "Then Hurricane Juan struck the area, which necessitated keeping some diving equipment in Halifax," relates Patrick, "and the low water prevented access to much of the area of interest. Overhead flights planned to see if the aircraft was visible—as was the case when the Search & Rescue technicians initially reported the sighting—were either canceled for weather or nothing of interest was seen. The divers saw nothing when they did dive, and so we began to question whether we were in the right lake, or the story of the Hudson was just that—a story."

Interest has not waned over the seven years of effort that have gone into searching for the elusive Hudson. Does it rest in a deep hole that the side-scan sonar was not able to penetrate? Are rocks blocking or reflecting

Divers investigate the crash of the Lockheed Hudson. (Courtesy Major Gen. (Ret'd.) Ian Patrick)

signals? Is heavy silt the culprit?

To date, divers in Gaspereau Lake have been rewarded with finding two civilian Cessnas and one tractor—but no Hudson. The History Channel, initially interested in obtaining the story of the Hudson, in the end did not come to Greenwood.

So, what's down there that can be the size of a Hudson—approximately forty-three feet in length, with a wingspan of sixty-five feet and an empty weight of 11,630 pounds? A whale? A boulder? Or the elusive Hudson, waiting for the right moment to emerge from the deep?

GHOST HUNTERS

On July 7, 1942, eight United States Army Air Force P-38s and two B-17s from the 94th Fighter Force squadron left from Presque Isle, Maine, on "Operation Bolero." Their flight path would take them via military air bases in Goose Bay (Labrador), Greenland, Iceland, and on to Scotland, where they would form part of a buildup of American warplanes in Great Britain. A P-38 leader and two wingmen escorted each fully crewed B-17, with the flights code-named "Tom Cat Green" and "Tom Cat Yellow."

Icing problems were experienced between Greenland and Iceland, so the squadrons laid over at Bluie-West 8 (BW-8) in Greenland. On July 14, the weather cleared and the ten aircraft took off for the six-and-one-half-hour flight to Iceland. Two P-38s soon encountered mechanical problems and turned back to BW-8, while the others, plus the two bombers, roared on through the icy air.

Heavy overcast was encountered even at twenty thousand feet. The squadrons tightened up and eventually lowered to fifteen hundred feet as they battled snow, rain, poor visibility, and low fuel. Even at this altitude they could barely see the rough seas and icebergs below.

When news was radioed that Reykjavik, Iceland, was weathered in and closed, the flight turned 180° to go back to BW-8. It too was closed. Bluie-West 1, an American search and rescue base on the southern tip of Greenland, was now their destination, but with low fuel the situation looked hopeless.

Colonel Joseph Bradley McManus, USAAC (Ret'd.) recalls that moment:

> After eight hours and five minutes of continuous flying, I called the flight leader and informed him that I was going downstairs. I peeled off and buzzed the icecap. The terrain was level and flat—concrete-like in appearance. I figured if I landed with my wheels down, I could save both the airplane and me . . . No one had ever said it couldn't be done, but then again, I would be the first one to try![1]

He lowered the gear and flaps, and throttled back.

When McManus brought in his P-38, the nose wheel collapsed, the front gear buckled, and the aircraft flipped and slid backwards to become buried under snow. The pilot escaped the smoke-filled cockpit by kicking through the Plexiglas canopy, then watched the other P-38s make successful gear-up landings nearby.

The two "Flying Fortress" B-17s, after using up their excess fuel and sending out SOS messages to anyone who might receive them, also made successful gear-up landings. All crew members were safe and uninjured, and prepared to camp on the solid ice. By cutting the prop blades off one of the B-17s, the engine was started to power the generator and charge the batteries so further messages could be sent.

When rescue came three days later by dog team, the crews learned they were seventeen miles from the east coast of Greenland. Before leaving, the airmen shot out the electronic equipment and instruments of their downed aircraft to prevent enemy retrieval. They were then taken to BW-1 via a Coast Guard ship and PBY Catalina aircraft, and home to the USA.

When the war ended and word of the downed aircraft began to circulate, people who were interested in scavenging parts came to find them. There were no planes. The epithet "The Lost Squadron" was coined.

Forty years later, in 1981, Americans Roy Shoffner and Pat Epps decided to form the Greenland Expedition Society and go in search of these warbirds. They had to be somewhere, under snow of course, and perhaps some ice, but certainly retrievable. But nothing was found, not even with metal detectors. Subsequent attempts failed, and five more years passed with funds and energy quickly dissipating.

By 1988, a plan was implemented to raise funds for retrieving the aircraft by selling one-eighth of a salvaged P-38 for twenty-five thousand dollars. With money in the bank, the society members returned with a subsurface radar device equipped to find objects buried in the ice. Back and forth went the crew, laboriously pulling the machine on a sled over the supposed graveyard of the P-38s and B-17s. Nothing registered.

They expanded their search area and were rewarded when the scope indicated an object below. A steam probe was inserted into the ice and hit something solid at the 250-foot depth. Unfortunately, bad weather delayed further investigations for another year, but in 1989 the team, led by Gordon Scott, returned with a core drill. The drill bit brought up a twisted piece of aluminum—they had found an aircraft. How could it be? A downed aircraft couldn't travel! But it had—over a mile in distance and 268 feet in depth through solid ice.

The final retrieval operation involved the use of a "Super Gopher," a four-foot-wide device supported by a steel frame that melts ice at the rate of two feet per hour by circulating hot water through copper tubing. It was a Herculean task, combining isolation, frustration, extremely hard work, high costs, and slow progress in biting winds and deep snow. When the Super Gopher had finally melted a four-foot shaft leading down to the object, team members were lowered into this dangerous passage by hand-cranked cables to inspect the findings. It was a B-17, totally crushed by ice, not repairable. There was no more money and no more energy left to continue.

Then, in 1992, Roy Shoffner contributed $250,000 for a further expedition, with Bob Cardin now the project manager. Following more than three weeks of back-breaking work, the Super Gopher reached the ultimate treasure: a P-38F Lightning. It seemed to be in good shape, so a decision was made to "melt" a cave around it to allow the crew members to dismantle what they could and attempt to bring up the aircraft.

The first items to reach the surface were four .50 caliber machine guns and the 20 mm cannon—all still operational. Next came the laborious task of bringing up engines and wing sections, and the huge twenty-one-foot-long center section that weighed over seven thousand pounds. The operation took eight days. By the end of the expedition, Roy Shoffner had spent six hundred thousand dollars to witness a helicopter transporting his treasure onto a ship and back to his hangar in Kentucky.

Then came the work—and funding—of restoration, mainly performed by Bob Cardin. Roy Shoffner, now the owner of the aircraft, had it painted in its original colors and appropriately named his prize *Glacier Girl*. On

October 26, 2002, Steve Hinton, considered one of the world's most experienced warbird pilots, performed the successful test flight before a crowd of twenty thousand people.

Glacier Girl, the only flying P-38F in the world out of the twelve thousand originally built during World War II, is now housed in the Lost Squadron Museum in Middlesboro, Kentucky. An impossible ten-year project, costing thousands of dollars, had come to a successful conclusion, bringing one of the Lost Squadron's aircraft up from the deep to fly the skies of southern USA.

In July 1998, the technicians who had recovered the P-38F in Greenland came to try their luck at resurrecting a USAF B-17 "Flying Fortress" (Bu. No. 44-83790) in Labrador. With only forty-three completely restored B-17s worldwide, and just fourteen operational, to find one intact underwater was an exciting prospect.

On Christmas Eve, 1947, the B-17 crew had been flying from Goose Bay, Labrador, to Greenland. When they became lost in a snowstorm and low on fuel, the crew had landed on the ice of an unknown lake. The Newfoundland base command searched one hundred thousand square miles of wilderness before finding the crew, all safe, thirty-six hours later on Lake Dyke, 270 miles northwest of Goose Bay—an amazing search and rescue feat. The heavy bomber had sunk through the ice.

With information that only the flaps and props of the B-17 had been damaged, the salvage technicians hired two local guides and set out to locate the aircraft from the known coordinates of its landing. Eight days later, they still had not found the B-17. Again and again the search area was widened until it was finally found—a full ten miles from the original crash site. The weight of the bomber had forced down the ice and in the spring it had been carried away by the current while resting on its icy platform twenty feet below the surface.

Diving and photographing operations ascertained that the B-17 was worth the cost of retrieval, but was the aircraft the property of the province of Newfoundland, the USAF, or the American salvagers? Major General Ian Patrick explains that, unlike in the ocean, where salvagers can claim rights, "inland waters are owned by the province, which sets its rules. The province claims rights to all artifacts in the water and seekers need a permit from the province to even look. The first permit allows one *only* to look. If anything is found, a report must be filed and then the province decides whether the seeker has the right to claim and raise the object(s). If the object is government-owned, permission from that agency is then required before the province will issue a salvage permit."

In the case of the Lockheed Hudson (whose story is related earlier), the aircraft belongs to Canada's Department of National Defence (DND). The closest base is usually charged with recovery if DND wants the aircraft, which it does. Foreign governments generally have first rights to anything that was theirs—such as the US government in the case of the B-17. If it gives up the right, or donates it, then recovery is possible by the seekers.

The B-17, in excellent shape from being preserved in the cold fresh water, is a treasure indeed. "When restored, it will fly again," says expedition member Robert Mester. And so, when the legalities are sorted out, the B-17 may join other underwater phantoms that rise from the deep to again embrace the sky.

BASHIN' BILL BARILKO

Bill Barilko was a handsome, hotshot hockey player whose short life became legendary, not only for his on-ice triumphs, but also for his tragic and untimely death under mysterious circumstances.

Born in Timmins, Ontario, on March 25, 1927, Barilko excelled at hockey from an early age, but it was not easy to make a successful career from the game. "Two dozen Timmins natives born between 1904 and 1946 played in the NHL," states author Douglas Hunter in *The Tim Horton Story*, "but more than half had league careers of 33 games or less. They were players who couldn't quite avoid the final cut . . . as Timmins' Bill Barilko put it, 'back to the arena with the leaky roof.'"[1]

Barilko was playing for California's Hollywood Wolves of the Pacific Coast Hockey League when he was called in February 1947 to join the Toronto Maple Leafs. "Hollywood" Bill Barilko instantly became a major player, securing a place on the National Hockey League's All-Star Roster for three consecutive years (1947, 1948, and 1949). On CBC's *Hockey Night in Canada*, announcer Foster Hewitt praised Barilko's energy and skill. Practically every family in the country who possessed a radio or television set tuned in to the program every Saturday night, and agreed with this assessment.

Bill Barilko's play showed energy and passion. The sturdy, five-foot-eleven-inch, 184-pound defenseman boasted a hard-hitting right-hand shot. His skating was fast and fearless, with strong body checks featuring darting "snake hips" maneuvers. Barilko and his defensive partner Garth Boesch would simultaneously drop to their knees to block offensive shots, a play that became known as their "Maginot Line," after the fortified defense line built along the eastern border of France following World War I.

The player who said he "couldn't skate, shoot or pass" became aggressive with "bashing" body checks and a related stack of penalties. "In his first full season with the Leafs he would rack up a league-leading 147 penalty minutes in 57 games while scoring five goals and assisting on nine," says Douglas Hunter.[2]

Although Barilko's scoring record was quite low, with just twenty-six goals in 252 league games and four in forty-six playoff games, the Toronto Maple Leafs #5 was closely watched by hockey fans—and other players. He'd certainly done his part to assist the Leafs in winning the Stanley Cup in 1947, 1948, and 1949. Now in 1951, they were in the playoffs once again.

THE GAME OF '51

The Toronto Maple Leafs and the Montreal Canadiens were the top teams vying for the Stanley Cup for the 1950–51 season. The Leafs had a 3–1 series lead, and each game in the finals had gone into overtime. The fifth game, played on home ice at Maple Leaf Gardens, was guaranteed to be dramatic.

In the first period, Bashin' Bill Barilko lived up to his nickname, garnering a penalty for charging, and in the third period for roughing. Just before the end of the third period, when the Canadiens were one point ahead, the Leafs removed their goalie. Leaf player Tod Sloan scored at 19:28 to bring about a 2–2 tie, sending the game once again into overtime.

At 2:53 in overtime, defenseman Barilko, showing amazing agility, leapt

Bill Barilko at the height of his hockey fame. (Courtesy Imperial Oil-Turofsky/Hockey Hall of Fame)

to stop "Rocket" Richard from scoring on the empty net. Barilko barreled down the ice from his spot at the left blue line, nearly colliding with his own center, Cal Gardner, grabbed a pass at the top of the left faceoff circle from Howie Meeker, and raced in to lift a backhanded shot past Gerry McNeil, the Canadiens' goalie. McNeil had been knocked off balance from two previous shots and never had a chance to stop it. Bashin' Bill Barilko slapped the winning goal into the upper right reaches of the net and instantly became a hockey legend. The Leafs had won the coveted Stanley Cup for the ninth time—and the fourth time in five years.[3]

But Barilko's forward movement didn't stop there. Accidentally catching the skate of teammate Cal Gardner, who was crossing from left to right in front of the net, Barilko went flying. This image, caught on tape, has become famous in the annals of hockey history. "He was out of position, out of turn, out of character, out of the blue," states author Douglas Hunter. "He could not have been more of a surprise had he jumped from the spectators onto the ice. His appearance and breathtaking success was a delightful discovery to the Leafs."[4]

Reporters swarmed the teams and their fans for quotes describing the phenomenon they'd all witnessed. The twenty-four-year-old defenseman had cause to celebrate: it was his fourth Stanley Cup since joining the Leafs four years earlier.

THE FISHING TRIP

Following the season's end, Barilko returned home to Timmins. In August, four months and five days after the famous goal, he decided to go fishing with a pilot friend, Dr. Albert Henry Hudson, who was a dentist. Barilko's mother and sister Ann Klisanich were uneasy about the trip. Why not go to a local lake if he wanted to fish? His mother pointed out that Bill didn't care much for eating fish anyway. Ann was concerned about the distance they'd be flying in a small aircraft over wilderness territory.

"Listen, Ann," was Bill's irritated response, "Mother is giving me a hard time and I don't know why." Then his tone softened. "Will you make a lunch for me?" Ann could only comply and wish him a good trip. Before leaving, Bill peeked into his mother's bedroom. "Goodbye, Mom, I'm leaving now." Still upset, she pretended to be asleep. "She never forgave herself for that," Ann said. "I believe she had a premonition something bad would happen and she didn't want him to go."

The two men took off in a single-engine float-equipped Fairchild 24w (CF-FXT) en route to Seal River, north of Fort George on James Bay. On the return trip, early on Sunday, August 26, Barilko and Hudson stopped at Fort George and inexplicably left their cooking utensils, tent, and sleeping bags with the Hudson's Bay Company manager. They declined an offer of a meal and continued on south for 160 miles to Rupert House, where they stopped to refuel three hours later.

Something seemed odd, thought Dan Wheeler, the Hudson's Bay Company employee at Rupert House, as he fueled the aircraft. It seemed unduly heavy. He decided to make a comment. "We had a great trip!" was the reply. "We've got about 125 fish stowed in the floats!" The weather was not good and they were encouraged to stay the night. "Thanks, but no thanks," was the cheerful response. "I'm used to flying in this weather." In actual fact, Hudson's logbook indicated less than three hundred hours flying time, considered inadequate by the military, as well as others who fly in such conditions, to qualify one to undertake a trip in poor weather in a low-powered and obviously heavily loaded aircraft.

They took off, with the aircraft taking an unusually long time to get airborne as it taxied back and forth and finally wobbled its way above the tree-tops. It never arrived at Timmins. When Barilko's family hadn't heard from him by that Sunday evening, they knew something was wrong. On Monday morning they received a call from a friend. "There might be a news bulletin on the radio. I don't want you to be alarmed, but it looks like something has happened to Henry Hudson's plane." He hastened to assure Ann that Dr. Hudson was known to be a good and conscientious pilot who was experienced at outdoor survival. They probably had to land somewhere because of the weather, or to fuel up or make a minor repair.

The media exploded with the news, and nothing could stop either the flow of factual information or the rumors that followed. Intense searches were conducted by both military and civilian search groups, to no avail. Strong headwinds had been reported in the area. An Austin Airways pilot, Captain Jim Hobbs, flying that day in the opposite direction—from South Porcupine to Moosonee—had made the trip in fifty minutes, less than half the time it usually took.

Friends and family, and the team and its fans, were devastated, and on

At 2:53 of the overtime, Bill Barilko scored the winning goal for the Toronto Maple Leafs, making them Stanley Cup winners for the fourth time in five years. (Courtesy Imperial Oil-Turofsky/Hockey Hall of Fame)

September 17 they posted a one-thousand-dollar reward for information on the lost Fairchild and its occupants. Conn Smythe, owner of the Toronto Maple Leafs, upped the ante by posting a ten-thousand-dollar reward. He also insisted on keeping Barilko's equipment hanging in his locker room stall, ready to go, should he show up for the 1951–52 season.

Finally, Smythe hired Tim Horton to take Barilko's place on the team. The men had uncanny similarities. Both were from the mining country of northern Ontario, Timmins and Cochrane, and both were aggressive play-ers. By that time, Horton had "collected 16 points and 85 minutes in penal-ties. In his final season, Barilko had 12 points and 96 penalty minutes."[5]

Smythe and others kept the faith that Barilko would be found unharmed, although general rumors called the disappearance a publicity stunt to enhance Barilko's status in the media. Others said that he'd "escaped" and returned to his ancestral land of Russia to teach hockey there. Further rumors connected the men with gold smuggling in the Timmins area, as Barilko's ambitions off-ice were known to be as strong as his hockey aspirations. Aside from the family appliance business, he and other Leafs members had interests in uranium and gold mining, and off-season time was largely spent up north.

"The Maple Leafs were shot through with mining fever," writes Douglas Hunter. "[James 'Gus'] Mortson and [Jimmy] Thomson were known as the 'Gold Dust Twins.' Barilko had his gold investments as well, as did Joe Primeau. Conn Smythe was ahead of everyone, having put money into gold in the mid-1920s with shares in Gold Hill Mining Co., and in the 1930s was a partner in Crossroads Gold Mines."[6] The report of heavy floats could mean that Hudson and Barilko had filled them with high-graded (stolen) ore concentrate or gold bars. The mystery intensified.

Searches were discontinued in November due to weather conditions, and resumed the following spring—and the year after that, and the year after that. Family and friends continued to hope for a successful conclusion. "Mother never, never gave up hope," said Ann. "From time to time she'd get me to drive her to fortune tellers, card readers, anyone she thought could tell her where she could find her son . . . Deep in mother's heart there was still hope. She wouldn't give up. 'A mother's wish is that her son will return some day,' she said."[7]

As years passed and the aircraft was not found, rumors continued to circulate of escapes to Russia and gold smuggling. Dr. Hudson was thought to be the fence who arranged transportation of the gold via his aircraft from the person who'd high-graded it, and was also suspected of being the one who would organize the eventual sale.

The Maple Leafs carried on, but seemed to be under a curse for they never won another Stanley Cup. Then, in the season of 1961–62, the Leafs defeated the Chicago Blackhawks to win the trophy, an honor that had evaded them since 1951, when Barilko had scored the never-to-be-forgotten, series-winning goal.

Almost simultaneously, the wreckage of the Fairchild 24 was found.

TRIUMPH AND TRAGEDY

In the last week of May 1962, Gary Fields, a helicopter pilot with the Ontario Department of Lands and Forests out of Cochrane, Ontario, and Ray Paterick, a passenger, noticed a flash of sunlight reflecting from a piece of metal. Fields mentioned it to co-workers, who, recalling the Fairchild mystery, took a closer look and recognized an aircraft wing and a float. Searchers returned to look for the ever-elusive wreck. Finally on June 6, the flashes of metal were seen again, and the location was marked by throwing out toilet paper that draped over the spruce trees. The wreckage was indeed the Fairchild, which had made a nose-dive into the bush forty-five air miles north of Cochrane, near Island Falls.

When investigators arrived at the site, they found a brush-covered, partly buried and burnt pile of scrap that had once been CF-FXT. The only items left basically intact were a fuel tank and an upended float. Inside the scorched cabin sat two skeletons, still strapped into their seats, identified as Hudson in the left and Barilko in the right. A picture of the wreck was featured in the *Toronto Star* on Friday, June 8, 1962, with an accompanying article:

COCHRANE (CP)

Wilderness only a few miles from one of Northern Ontario's transportation lifelines has ended its 11-year concealment of the fate of Bill Barilko and Dr. Henry Hudson. Searchers yesterday found the skeletons of the two men in the smashed wreckage of their plane, which apparently caught fire in crashing 45 air miles north of here and a few miles east of the Ontario Northland Railway. Searchers were led to the scene by a helicopter crew of the Ontario Department of Lands and Forests, who spotted a glint of metal in the dense bushland (from the airplane which had) disappeared in 1951.

The Department of Transport's air accident report noted that "the aircraft had struck the trees in a 50° nose-down attitude at an angle of bank of 25° to the left. Damage to the propeller indicated that the engine had been either underpower at the time or windmilling at high rpm. The elevator trim tab was found in the almost full nose-up position. The trim control in the cockpit corresponded to this position. There is no evidence available to explain this trim setting, which is considered abnormal to cruising flight. There is insufficient evidence to determine the cause of the accident."[8]

Could it have been so trimmed because the aircraft hadn't been properly loaded? Had the center of gravity been offset by a heavy object being carried near the front, or the extraordinary weight in the floats? There was no evidence of anyone having tampered with the wreckage, but the inordinate police activity that surrounded the site caused rumors to proliferate and indicated that "this was no ordinary investigation."[9]

The floats and their contents—whether water, fish, or the rumored gold—seemed to be of special interest. Fish bones are soft and deteriorate quickly, and if the floats had split on impact, animals could have cleaned them out. But what about the intact float? Without elaborating further, Corporal G. Duguid of the Ontario Provincial Police stated at the inquest that there was no evidence of fish or fish bones found in the floats or in the wreckage.[10]

If there was gold bullion in the float compartments—and such stories continued to circulate—it would not have deteriorated over the years. Is that why the police had so carefully guarded the wreckage? Peter Worthington, reporting for the *Toronto Telegram* on June 7, 1962, acknowledged that "although the bush has surrendered the victims, the mystery still lingers." What had caused the plane to crash? It surely couldn't have run out of fuel, with two full tanks holding thirty US gallons each. Ontario Provincial Air Services personnel later found Hudson's fuel caches at various sites on the east coast of James Bay; he probably also had caches between Rupert House and South Porcupine, although these remained undiscovered. The accident could have been influenced by the bad weather and turbulence, combined with a relatively inexperienced pilot and an aircraft underpowered for its load. But why was the wrecked aircraft found to be pointing north when the flight was due south?

No specific cause for the accident could be determined by the department of transport investigation. No public statement was issued on the reason for the extraordinarily long and thorough investigation of the crash site conducted by the Ontario Provincial Police before the bodies were allowed to be removed. The mystery remains as intact as the upended float.

Jim McAvoy, a Canadian bush pilot who is familiar with both this type of aircraft and with gold mining operations, believes the idea of gold smuggling is absurd. "For one thing," he says, "gold is hard to get hold of, and harder to sell. There was likely water in the floats. The aircraft was probably overloaded for its power, that's why they had trouble getting into the air off the lake."

Those of a superstitious nature noted the frequency of the number five: it was Barilko's jersey number, he'd scored goal number five in the fifth game of the '51 series, and helicopter pilot Gary Fields was five days into his summer contract when he'd discovered the wreckage.

Fellow players remember the shock that resonated throughout the hockey world at the discovery. "We were just driving away from the Gardens [after a team function] when someone ran up to me and said, 'They've found Barilko's body,'" Frank Mahovlich recalls. "I was stunned. Everyone on the team made the connection between Bill's goal in 1951 and the fact we hadn't won until then."[11]

Was this a coincidence? A sequence of events controlled by a mystical power? Barilko's spirit returning to his team? The stories still linger.

THE RETIRED #5 JERSEY

The Toronto Maple Leafs "retire" the number of a player who has made significant contributions to the team only when his career has been cut short by injury or death. These numbers can never be used by any other player. Just two retired numbers celebrate Leafs players: Bill Barilko's number five and Ace Bailey's number six (Gretzky's number ninety-nine is an anomaly, being retired leaguewide by the NHL). These uniforms, in the Leafs' road colors of blue with white trim, were suspended from the rafters of Maple Leaf Gardens and are now in the Air Canada Centre, the home ice of the Toronto Maple Leafs.

And so the legend of Bashin' Bill Barilko remains a source of pride, sorrow, and mystery. Leafs fans view the retired jersey at each home game and remember, or are told about, the phenomenal young player from Timmins. Emotions are again stirred by the song, "Fifty Mission Cap," by the Tragically Hip, played during every home game, which recalls Barilko's famous goal and tragic disappearance.[12] Canada's Hockey Hall of Fame in Hamilton, Ontario, displays the sweaters worn by Barilko and goalie Gerry McNeil, as well as the winning puck from the 1951 game.

Canadian art curator Andrew Hunter wrote a novel and based an exhibition on the story. Titled *Up North, A Northern Ontario Tragedy*, the work considers a link between Group of Seven painter Tom Thomson and Bill Barilko, who both died in the wilderness of northern Ontario. Blending historical fact with myth, Hunter envisions a spiritual connection between the men and their disappearances.

Bill Barilko's fate also shares similarities with that of Tim Horton: "Barilko, a defenceman from Timmins, won four Stanley Cups—three in a row, then a fourth after a one-season dry spot—before dying in a plane crash . . . Horton, a defenceman from just up the TNO (the Ontario government's Temiskaming and Northern Ontario Railway, later Ontario Northland) railway line in Cochrane, was brought in to replace him. He too won four

Stanley Cups as a Maple Leaf—three in a row, then a fourth after a two-season dry spot—before dying in a car crash on February 21, 1974."[13]

For the Stanley Cup playoffs in 2001, the fiftieth anniversary of that final game and subsequent death of their legendary defenseman, the Leafs featured Barilko's picture on the cover of their postseason media guide. It was obvious that fans still remembered him, as did fellow hockey players. Expressions such as "fearless," "had no peer," and "one of the greatest defensemen" were used by many. Maurice Richard recalled Barilko's toughness, in both body and spirit, that earned him his share of penalties for roughing. Gerry McNeil called him a hard worker. Team member Howie Meeker noted that for the defenseman to be "as deep as he was, at that point of the game, was dangerous—but that was Bill Barilko. It was not a surprise. He was tough. When he did body-checks he really hurt people. It was like being hit with an end of a pick-ax or shovel."[14]

Barilko's sister, Ann Klisanich, took part in a ceremony honoring her late brother at the Air Canada Centre in the 2001 home playoff game, during which CBC's *Hockey Night in Canada* and other media presented tributes. Included was film footage of the "famous goal of '51" featuring Barilko soaring toward the net after taking the puck from the top of the faceoff circle.

A poignant article appeared in the *National Post* on December 11, 2004, that brought the Barilko tragedy to readers' attention once again. Reporter Joe O'Connor's interview with Louise Hastings of Toronto revealed the heartache still felt by Barilko's sweetheart over his loss. He had bought a ring, Ms. Hastings said, with the intention of presenting it to her on her birthday that September 4. His aircraft went missing just ten days earlier, on August 26. She still grieves, and, over fifty years later, keeps his photograph by her bedside.

Scrapbooks and albums owned by Barilko's family, friends, and fiancée contain copies of newspaper coverage chronicling the disappearance of the aircraft and discovery of the bodies. But there are sweeter memories of this enigmatic young man, Bashin' Bill Barilko, and those who knew him hold onto them with reverence.

BRITISH COLUMBIA'S
MISSING IN ACTION FILES

Canada's most westerly province has rightfully earned the macabre name, "Graveyard of Lost Planes." British Columbia is a wild and beautiful part of the country, but not a place for inexperienced pilots—or perhaps even experienced ones. The climate varies dramatically across the province, from rain, coastal sleet, and high ocean waves to blinding, whiteout snowstorms, and even blistering heat in the southern interior. British Columbia's windswept mountains and vast wildernesses are covered by trees, pocked by bottomless lakes, and sliced by rivers and rocky crevasses.

Flying over mountain ranges provides the ultimate challenge to pilots. This is evidenced by the number of aircraft that rest in crags and valleys, remain frozen in and beneath glaciers, lie broken and buried in overburden at the bases of avalanches, or nestle among trees or under brush.

"Disappeared without a trace" is a phrase too often read in newspapers or magazine articles. On July 21, 1951, the *Vancouver Daily Province* noted that "in the last nine years, some 112 persons have vanished in 16 planes." *Macleans* magazine followed on January 4, 1958, with an article titled "The Toughest Flying Country in the World," where Ray Gardner asserts that no other region in Canada, and few in the world, had brought down more aircraft than British Columbia. In a sixteen-year period from 1941 to 1958, 117 lives had been lost in fifteen fatal crashes involving both small and large aircraft.

On October 17, 1976, the *Victoria Times* discussed the phenomenon of BC's lost planes. "Over the past 20 years, 97 people vanished from BC skies. Up to February 1972, none of the lost aircraft carried emergency locator transmitters, devices that broadcast a signal in the event of a crash," acknowledges reporter Ab Kent, but adds, "Since then, four planes equipped with ELTs went down without being found . . . so even with such improvements, planes still are swallowed by the sea or snatched by one of a thousand rocky crags or ice crevasses spanning much of the province."

The grim count of fatalities has continued into the twenty-first century. In 2003 alone, air accidents took the lives of a young pilot who'd taken up skydivers near Chilliwack in June; three occupants of a Cessna 172 on Willie's Peak near Golden in July; two pilots in a four-engine Lockheed Electra aerial tanker fighting forest fires near Cranbrook in July; three people in a deHavilland Beaver flying from Penticton to Calgary in August; and a pilot in a Cessna 414 flying from Cranbrook to Calgary in September.

While there are fewer "mysterious" disappearances or unfound crash sites in recent years with the advent of radar coverage and electronic search devices, the British Columbia wilderness still offers a multitude of hiding places. Some remain undiscovered for years until a hiker, geologist, or bush pilot spots something odd and decides to check it out.

DEATH FROM ABOVE

On February 13, 1950, a crash that still evokes a sense of danger occurred in the Kispiox range of the Coast Mountains of northwestern British Columbia (thirty miles northwest of Hazelton). Kologet Mountain became the repository for a stray USAF Convair B-36B "Peacemaker" strategic bomber, reportedly worth six million dollars. The B-36 was the largest bomber ever built in the USA, with a wingspan of 230 feet, and powered by six engines with pusher propellers that faced backward rather than forward.

The training mission (Flight #2075) left Eielsen Air Force Base at Fairbanks, Alaska, to fly south to California, then to its home base at Carswell Air Force Base, Fort Worth, Texas. The objective was to reach the "lower 49" states and then do "supposed bombing" from Los Angeles to Texas. When the huge bomber picked up ice while flying over northern BC, it began to lose altitude at the rate of three hundred feet per minute. Then fires were indicated in engines one, two, and five, and a plugged gas line shut down number three.

At eight thousand feet, the bomber jettisoned its cargo—an atomic bomb—over Hecate Strait. The eleven-thousand-pound Mark IV, "Fat Man" A-bomb—similar to the one dropped on Nagasaki, Japan—apparently exploded over the water, but because its plutonium capsule was not installed, it was not a nuclear detonation. Most of the thirty-six detonators on board supposedly went down with the bomb, but some remained in the aircraft along with 252 rounds of 20-mm ammunition for the bomber's anti-aircraft guns. This was the first time the United States had "lost" a nuclear weapon, and it happened over British Columbia. The USAF code-named this accident "Broken Arrow," signifying that a nuclear weapon had failed to meet its intended mark.

The seventeen people aboard—sixteen military crew and one passenger—from USAF 436th Bomb Squadron, 7th Bomb Group, bailed out over Princess Royal Island, south of Prince Rupert, supposedly at three thousand feet. The aircraft, with three of its six engines reported out of commission, then flew 223 miles further before crashing near the six-thousand-foot level of Kologet Mountain.

Twelve of those who bailed out were found—one dangled upside down by his parachute strap from a tree branch for twelve hours before rescue—but five others vanished without a trace. The aircraft remained

The giant XB-36 shown here dwarfs a Boeing B-29. (Courtesy Phillip Jarrett)

undiscovered for three years until 1953, when the RCAF found it while searching for another missing aircraft. That fall, a group of USAF investigators attempted to check out the site, but were prevented from reaching it due to the difficult terrain and early snow.

With the tensions of the Cold War heightening in 1954, US engineers could wait no longer. They rushed back to Canada to salvage valuable electronic materials from the Broken Arrow crash site and then blew up the wrecked B-36 to prevent its secrets from falling into enemy hands.

This was not the end of the Broken Arrow episode. Further evidence of the wreckage was discovered a few years later by a mapping crew, but it was not until 1996 that a former employee of the Geological Survey of Canada reported the discovery to the Environmental Protection Branch, Pacific & Yukon Region, Environment Canada. Back in 1956, while doing survey work on Kologet Mountain, he and a coworker found an unopened canister with USAF markings attached to a parachute. Inside was a Geiger counter. The aircraft might indeed have been carrying a nuclear device, which could have left radioactive material at the site. Canada's Department of National Defence was contacted and in August 1997 they arrived to conduct a field assessment.

The tail section, port wing, and three engines were found scattered along a steep west-facing slope; further away were remains of the starboard wing, engines, gun turrets, and main fuselage, which still contained some personal belongings. Fire and explosives had destroyed portions of the aircraft. Four electronic detonators were found, with the remaining thirty-two detonators possibly used to arm the bomb that was released before the crash. As well, there were canisters of unused explosives attached to parachutes, likely from the USAF's visits in 1953–54 to destroy the bomber. Ironically, the insignia from a 7th Bomb Wing pin found among the wreckage was inscribed with their motto, *Mors Ab Alto*—Death from Above.

A number of questions about the bomber remain unanswered and might stay that way. Many people with knowledge of the incident are now dead, and those still alive steadfastly refuse to talk about it—"or they *will* be dead," ventures a retired RCAF pilot. How did the aircraft fly over two hundred miles, unmanned? If the crew bailed out at three thousand feet, how could the aircraft have hit the mountain at six thousand feet? (However, reports vary. Captain Harold L. Barry reported they were heading 165° at five thousand feet when the men bailed out, ten seconds apart).

If the aircraft was heading south when it ran into trouble along the coast, what caused it to turn inland before crashing? The explanation that it had been set on automatic pilot doesn't answer that question, nor does speculation that the engines somehow restarted after the crew had bailed out, and due to an error in the automatic pilot, the aircraft circled in an arc to fly to the northeast. Could one of the captains have stayed with his ship to try to steer a course? Do his bones rest with the wreckage on the mountain, or were they removed when the USAF arrived to destroy evidence?

Was a plutonium core for the bomb aboard the B-36? Did the crew bail out with it, or jettison it as well? Did US engineers retrieve it in 1954? Or could it still be among the debris on the mountainside, undetected in its shielded case?

British Columbia writer Dirk Septer has a special interest in this mystery. Satisfactory answers to his queries have not been forthcoming, however, even after "endless correspondence with the Air Force Historical Research Agency at the Maxwell AFB and the US Total Army Personnel Command." Septer notes that "a trail of misinformation surrounds the incident. If this was intentional, these efforts have been quite successful."[1] "This bomber's got a story to tell and it hasn't been told—it's been kept a secret for so long," Carl Healey of the Broken Arrow Aircraft Society stated to *Terrace Standard* reporter Jeff Nagel, during a site examination in 2000.

Over the years, visitors have helped themselves to souvenirs, including rounds of 20-mm ammunition for the bomber's anti-aircraft guns, cannon turrets, aircraft insignia, and personal items. One souvenir hunter found a detonator in the rubble and carted it home. When the RCMP discovered its whereabouts, the volatile item was immediately confiscated. "Had it detonated, it could have flattened the man's home," said Master Corporal Pierre Coté, Canadian Armed Forces Rangers, Terrace, BC, who is also a member of the local Broken Arrow Aircraft Society.[2]

Because crash sites more than two years old are considered heritage sites, it is forbidden to remove artifacts from the area. This is discouraging to organizations such as the Broken Arrow Aircraft Society, whose members hope to preserve the artifacts for a museum in Terrace, and the Stewart Historical Society Museum in Stewart, BC, which has in its possession a cannon barrel, apparel, and equipment.

Although Environment Canada announced that little radiation danger remained, the only sources being the radium-illuminated electronics and gauges, the crash site was checked again in October 2002. The purpose was to make a thorough search for any hazardous materials, including small gold and stainless steel detonators, and get rid of them once and for all. An Explosive Ordnance Demolition team from the Canadian Armed Forces Base at Wainwright, Alberta, along with two Griffon military helicopters, assisted with the work, as did Canadian Armed Forces Rangers from northwestern BC and the 4th Canadian Ranger Patrol Group from Victoria.

They found more explosives littering the mountain than they'd even imagined, including those left behind in 1954 by the US Army Engineers who came to demolish the wreckage. The CAF teams destroyed a one-hundred-pound canister of highly explosive Compound "B." Their metal detector also located 9-mm and 20-mm cannon shells with loose explosives and explosive tips from the bomber's anti-aircraft guns, and meters of detonator cord. In all, 250 pounds of explosive materials were found by the team. No more detonators were found. "It's believed that most detonators went down with the bomb," states Jeff Nagel in a "Special to the News" report, "but Master Corporal Pierre Coté said there are two that are unaccounted for and could still be on the mountain."[3] Coté had earlier expressed the view that with the Soviet Union having detonated its first atomic bomb just six months before the crash, heightening the Cold War tensions of the time, "there was *no way* they were flying without a live core."[4]

The location of the wreckage in the bowl of a broad alpine valley surrounded by high ridges made it appear that it was a planned landing. On October 16, 2002, Coté reported that his group was informed that a body had been removed from the wreckage in 1954, adding his opinion that "there's no way the plane got to where it got without someone piloting it in there."[5]

On November 3, 2002, the Explosive Ordnance Demolition team helicoptered back to the site along with 120 pounds of C-4 plastic explosive to detonate the dangerous findings. "What they lit off there was enough to get rid of most of Kitimat," reported the leader of the 4th Canadian Ranger Patrol Group, Kitimat. "We were 12 miles away in the base camp and we heard the boom."[6] More information may be revealed in 2010 and 2015, when further US military documents are due to be declassified.

A documentary film, *Lost Nuke*, was made in 2004 to later air on the Discovery Channel. Film sequences were shot off the waters of Prince Rupert, BC, and in a special set constructed in a hangar in Edmonton, Alberta, to tell the story of the world's first lost nuclear weapon. "Attention George W. Bush," announced *Terrace Standard* headlines when describing the film project. "If you're looking for weapons of mass destruction, try Edmonton."[7]

——— ——— ———

THE HAUNTED ISLAND

The love of flying can become an obsession, and so it was with Edward Hadgkiss after he moved from BC's Lower Mainland to Whitehorse, Yukon. He took flying lessons there and when he returned home for a visit he bought a Cessna 120. On his way back to Whitehorse he had to make a forced landing on the highway due to a mechanical failure in the newly serviced engine—strike one.

His next acquisition was a complete change from the little Cessna: a Harvard Mark IV (CF-XEN), principally used in training World War II fighter pilots. He named it *The Yellow Peril*.

Ed's flying experience was anything but uneventful. On a practice touch and go in a training aircraft in Whitehorse, he was informed by air traffic control that his right landing gear had broken and was swinging around beneath the aircraft. He did another circuit. As his airspeed slowed and he was touching down, he kicked in full opposite rudder to ground-loop the aircraft away from the broken gear. The landing was successful, with just a few scrapes to the aircraft—strike two.

In November 1969, Ed and his girlfriend, Katherin Rheaume, decided

Ed Hadgkiss with his Harvard at Haines, Alaska, August 1968. (Courtesy Jane Gaffin)

to take a trip from Whitehorse south to San Francisco. When storms delayed their departure for three days, they knew they would be flying through a notorious weather area along BC's west coast, and in the most volatile season. Hadgkiss had only a VFR rating, decreasing the margin of safety, but they went ahead with their plan.

Refueling stops were scheduled for Juneau, Alaska; Prince Rupert, BC; Port Hardy on Vancouver Island; and Pitt Meadows in BC's Lower Mainland. But their landing in Port Hardy was stymied by weather, with fog totally obscuring the coast. Ed turned the Harvard north to retrace the seventy-five-minute flight back to Prince Rupert. Sheets of freezing rain pelted the aircraft. He turned east toward the mountains, hoping to find one of several small airstrips that he knew existed along the coast. When he spotted a ridge of land covered with small trees, he came in for a landing at slightly less than one hundred miles per hour. He was prepared for the worst, and got it—strike three.

The Harvard's tail reared up, it flipped over and skidded downhill on its back for one thousand feet. It finally halted on the edge of a ravine, braked

The Harvard Mark IV (CF–XEN) came to a rest upside down and intact in a clump of fir trees on Roderick Island. (Courtesy Jane Gaffin)

Ed and Kathy's camp, June 1970. The circle indicates silk material found clinging to the bushes, showing where the survivors had created a shelter out of a parachute. The horizontal line indicates the snow level when the camp was found in February 1970. (Courtesy Jane Gaffin)

by a grove of fir and hemlock trees. Ed and Katherin crawled out of the wreckage and surveyed the damage. They had survival gear and had filed a flight plan, so had every hope of being rescued. Ed's last note in the aircraft logbook reads, "November 10, 1969, 13:00 from Prince Rupert to 14:00 crashed on top of a mountain."[1] What happened after that remains a mystery.

Severe weather conditions and radio signals whose source couldn't be pinpointed hampered searchers. Finally, on February 22, 1970, the Harvard was located on the "dark and foreboding" Roderick Island, an uninhabited place fearfully avoided by the Klemtu First Nation, who lived at the fishing village of Klemtu on nearby Swindle Island.

"According to legends," writes Tessa Derksen in *Aviator Magazine,* "the Red Sasquatch—a mythical creature that carried off naughty children— haunted the island and people had disappeared before, swallowed by Roderick Island."[2] Klemtu people questioned by searchers about Roderick Island confirmed they'd just recently seen a six-foot-tall Sasquatch standing on the shore, and didn't want to go anywhere near the place.

The Harvard was found mostly intact, though inverted, with some parts and cargo littering the snow-covered rocks of the diabolical island. Nearby were traces of a camp apparently used by Ed and Kathy, who'd tidied it up before leaving the site. Searchers found Polaroid pictures they'd taken of each other, some in front of the parachute tent they'd hung over a tree branch, others while wrapped in a sleeping bag, smiling for the camera. Nothing indicated injury or fear. But where were they now?

Ed Hadgkiss's logbook was found, with notes describing the situation, as well as a letter dated November 18, 1969, that outlined their plan:

> After staying here for 7 days in very poor weather, today we could see the ocean to the west about 10 miles. We saw a tug pulling a barge and know there is a lighthouse there somewhere 'cause we could hear it when it is foggy. Myself and Katherine

[sic] Rheaume, having not been hurt in the wreck, have decided to take what we can carry and walk towards the ocean west of here. We have at least 4 days food left, flare guns, .22 rifle, a parachute that we have been using as a tent and some gas and oil to help lighting a fire. We intend to follow the ridge west of here fore [sic] the first while, then either left or right of the knob and then down to the sea.

The aircraft was not insured and I as the owner do not intend to do anything further with it. I have removed some of the gauges hoping that we could ride out in a chopper but they are here still with tools, Coleman stove, etc., that were too heavy to pack. Anyone visiting this site is welcome to anything here. I feel that we were very fortunate to escape with our lives.

I would appreciate to hear from anyone who has been here. Perhaps you could mail me a souvenir as all I have to show for an airplane I was very fond of is a timepiece and compass.

Edward J. Hadgkiss
Box 2474, Whitehorse, Yukon

Nearby residents said that around Christmastime they'd seen a flare from the island's west coast on Finlayson Channel. Someone found a deer hide near a river at Mary's Cove at the west end of Roderick Island; investigators confirmed it had been killed by a human hunter. Could it have been Ed Hadgkiss, who had guns and ammunition with him on the aircraft?

Searchers combed the island, slogging through swamp and bog, and narrowly escaping falls over obscured cliff edges as they chopped paths through the dense evergreen salal. Exhausted, wet, and cold to the bone, they now understood the basis for the Klemtu legends: those who go to Roderick Island never return.

Numerous pieces of evidence were found as far away as three miles from the wreckage: expended 22 rifle shells and .455 cartridges from a Webley handgun still registered in Ed Hadgkiss's name, a man's boot and a woman's tote bag (near a small lake on the island), and a gum wrapper (near Jackson Lake). Then Ed's club bag was found at the base of a cliff, containing a soapstone carving he was bringing to his parents, men's and women's underclothes, and the Harvard clock. Fifteen feet up in a tree, searchers found Ed's wallet and rolled-up socks. At the base of a cliff, they found Kathy's leather boots wrapped in a torn sleeping bag.

Had the couple fallen? Both east and west sides of the tree-covered ridge leading down from the wreckage were combed, which took searchers, after a three-hour walk, to Mathieson Channel. When nothing further was found, the search was abandoned on June 20, 1970, and personal effects were returned to the families with letters of condolence.

It was concluded that Ed and Kathy had perished by falling from a cliff and hitting trees or a ledge. Their bodies had likely been buried by debris or destroyed by animals. However, some people disagree. If they'd fallen down a steep slope or cliff, why weren't their bodies found near the personal belongings they were carrying? Search dogs surely would have found them had they been buried in snow or overburden, and the predatory animals of Roderick Island, such as small wolves, would not bother with dead bodies when deer were so abundant.

Doug Banks of Tofino, BC, a pilot familiar with the wilderness, developed an interest in this case and has hiked to the crash site twice. "People said it was impossible to get to," says Doug, "but ten years after the accident I flew over it in a Beaver and saw the best way to get up there. So in March I hiked up to the four-thousand-foot level. There was some snow at the lower level and by the time I got to the top it was up to my knees." The first time there he spotted a number of clues that had been missed by initial searchers. "They should have sent a bushman in at first to track them," he said, "but the hoards of people who helped in the search actually contaminated the site."

Teeth marks on .22 bullets showed where someone had bitten out the lead slug to use the gunpowder to get a fire going. A partly burnt piece of the blind flying hood from the aircraft and a burnt corner of a map had been used to fuel their campfire. "I wish they'd gone to Mexico like some believed," Doug says, "but from the evidence I'd say they got hypothermia. It's a really bleak part of the coast. You get cold and wet in that country and with only a poor fire, or none at all, you never get dry. They stayed on the hogsback for a while, then likely took shelter under a ledge and died there. Or perhaps they heard boats and tried to walk down and slipped on the rocks."

On his second trip to the crash site, Doug Banks noticed that some instruments and other parts had been taken from the Harvard. Although he declares a respect for aircraft crash sites and hates to see downed aircraft stripped of parts by souvenir hunters, he felt compelled to take a part himself, the control column, which he later gave to a friend in Vancouver. The haunted legend of the island again sprang to life.

Dr. John (Jack) Albrecht of Burnaby, BC, was the recipient of this gift, which he had planned to fashion into a lamp stand. His house has been "haunted" ever since. "By the year 2004 we will have lived in this house for twenty years," Dr. Albrecht says. "After I brought the Harvard control column home in 2000, spooky things started happening. I checked with the two previous owners of the house, and nothing like this had ever happened when they were here."

The occurrences began on Thanksgiving Day, 2000, when he, his wife, and two daughters suddenly heard a bizarre "whoa-ooh" sound that rose and fell. They investigated every room, the carport, the houses next door, but couldn't find the source. The sound was loud and eerie, with the tone and cadence of a foghorn.

"I'd been nervous about a proposed sailing trip we were booked to go on," Dr. Albrecht says. "Three days before our flight to the Caribbean, I had two sequential dreams in one night. In the first I was drowning, but suddenly my lips were being pulled apart and cold air being forced into my mouth. I woke up in a sweat, still feeling the cold air pursing my lips and mouth. In the second dream I was climbing the stairs of the hospital and had reached the sixth floor when suddenly sparkling, silver lights surrounded me. I woke up in bed, literally paralyzed. I couldn't move my arms or legs. A voice, with the same deep cadence as that I'd heard two months before, was saying, 'Don't go!' Then the sparkles disappeared and I could move again."

Jack and his wife, Ruth, discussed the dreams, the sounds, and the trip. "Should we go?" Ruth asked. He thought for a moment then decided. "Yes." Their trip to the British Virgin Islands during the first weekend of November 2000 became sandwiched between a tropical storm and a hurricane that tore down buildings and wrecked ships both at sea and in the marinas. "We happened to arrive and be on the water exactly between these two storms," Jack says. "We had minor adventures, but were kept safe somehow. Then, flying back from Costa Rica, another jetliner crossed directly in front of our aircraft, at exactly the same altitude! I could see the vortices as our flight paths crossed. There was no evasive action taken. I have never witnessed such a close call to a midair collision."

The weird occurrences have continued; objects are frequently moved about. One night, the fireplace implements suddenly began swinging like pendulums. "When I tried to simulate the maneuvers, it was impossible; the implements simply would not move in that pattern. At times I'll be sitting in our living room and feel a puff of cold air brush past my face," he says.

Even though these strange phenomena still go on in his home, Dr. Albrecht says he has no plans to part with the Harvard control column. "The last person who touched this column was Edward Hadgkiss, and rarely a day goes by that I don't think of him and his girlfriend Kathy Rheaume," he says. "I think somehow that their spirits reside in this house, they came along with the part from the Harvard. But they are friendly spirits sent to look after us, to warn us of possible danger." He adds, "A company is now logging on Roderick Island, so someday perhaps they'll find the remains of these two young people, or find a telling clue to help solve the

Dr. Jack Albrecht holds the "haunted" Harvard control column. (Courtesy Dr. Jack Albrecht)

mystery of what happened to them."

"Yes, the island is believed to be haunted," Doug Banks agrees. "That didn't help in the search because the Native people, who know the land, would have nothing to do with it. Then I gave the stick to Jack and he says his house is now haunted. I don't know what happened to those who took the other Harvard parts. It would be interesting to find out."

Wilderness Airlines of Bella Coola, BC, was given the salvage rights to the Harvard. In 1986, seventeen years after the accident, the aircraft was dismantled and airlifted by helicopter to Shearwater, near Bella Bella, and then barged to Vancouver. Roderick Island continues to guard its secrets. Perhaps, as logging operations plunge deep into its forests, the legend of the Red Sasquatch will be resolved and the spirits of those who lost their lives on this dark island will be freed.

THE LOST BOYS

On April 29, 1982, a Cessna 185 II (C-GJLJ) went missing in the rugged Monkman Pass between Fox Creek, Alberta, 155 miles northwest of Edmonton, and Prince George, BC. The recently purchased aircraft, white with an orange stripe, carried five men: Rick Gascon, George Maurer, Larry Anderson, and Darryl and Brian Trottier. The Trottier boys were brothers of Dale "Trapper" Trottier, seven-time Canadian bareback rodeo champion, and all men aboard the Cessna were involved in the oilfield trucking business. Their destination was a heavy equipment auction in Prince George, 250 miles from Fox Creek.

When the missing report was issued, "a brotherhood of cowboys" assisted Dale Trottier to form the Rocky Mountain Air Search Association. Help also came from military and civilian flyers, and personal friends such as Alaska Highway–based pilot Jimmy "Midnight" Anderson, who logged more than four hundred flying hours in the search operation. The official search was conducted over the twenty-one-thousand-square-mile area between Grande Prairie and Prince George, 85 percent of which was mountain country, at three altitude levels. The Canadian Forces Search and Rescue teams, along with fifty-seven civilian planes, were registered to help with the search.

Kathlyn Rhea, a veteran psychic of various FBI and police investigations in the USA and the author of two books, came from California to meet the families and to fly the route. She sensed that the aircraft had completed one-third of its journey, then gone out of control and crashed somewhere near the Kakwa River or Musreau Lake vicinity (13.5 nautical miles north-northwest of Smoky Tower Forestry Airstrip), leaving no survivors.

Donald Abbott agrees that "this area is prime." He was the principal western civil aircraft accident investigator for Transport Canada assigned to the case, along with his partner, engineer Bill Rahn. They listened to the last voice communications of Darryl Trottier, the Cessna's pilot, as he spoke on

A typical Cessna 185. (Courtesy Aero Space Museum Association of Calgary and Art Smith Aero Centre)

April 29, 1982, to Grande Prairie Flight Services on 126.7 at 1348:18z. The flight reported at ten thousand feet over Smoky Tower—a forest-fire lookout station located fifty-four nautical miles along the track of JLJ's flight path from Fox Creek to Prince George. The transmission lasted for seven seconds— "Juliet Lima Juliet over Smoky Tower Ten Thousand Feet [mumble] . . ." and then ended.

As Abbott listened to the recording, he heard a microphone click at 1314:13z, followed by background noise from the open mike. He says, "This signified that weather might have been so turbulent that his hand had slipped off the transmitter switch. If there is moderate to severe turbulence you must hold your hands secure or they, and any loose components, will fly about."

The upper winds on that day would have caused severe turbulence, which a pilot would not be able to anticipate unless he had extensive mountain experience, Abbott says. Darryl Trottier had by that time accumulated three hundred hours of flight experience, of which a portion would have been into mountainous areas. Abbott adds, "And as a chap who had worked in the mountains with heavy trucks, he was certainly knowledgeable of how the weather could suddenly change and that winds could be unpredictable at the best of times."

Aircraft that crash in these circumstances are often difficult to find because of the extreme high speed and disintegration on impact. Although Abbott has never heard of a Cessna 185 breaking up in flight, he did look into this possibility at the time. He further searched in the "area of most probability" from a deHavilland Twin Otter and also from his own C-185.

Donald Abbott and his partner had the sad duty of explaining to the families of those aboard the missing Cessna C-GJLJ why it might be so difficult to find the downed aircraft, which could have contacted the terrain at speeds in excess of 170 miles per hour. "Darryl Trottier's wife, Audrey, as well as the other four wives and children of the missing men, were in attendance. I recall the wives being as one in their resolve to not give up hope, yet at the same time they were realistic of the burden they faced together."

On May 13, 1982, *Edmonton Journal* reporter Tom Barrett spoke to the families involved. Two of the men's wives had pleaded with their husbands not to leave that morning. Laural Gascon remembered it being terribly windy: "As Rick was leaving I told him, 'If you go I know you won't be home.'" Audrey and Darryl Trottier had an argument about the flight. "'It wasn't the weather, it was just a feeling,' she recalled. 'I got downright mad at him. I hate that plane.'" "In fact," the reporter concluded, "the women have jokingly planned to make a huge bonfire of the plane when it's discovered. Their only wish is that their husbands will be with them to watch it burn."

Almost ten years later, on November 17, 1991, the *Edmonton Sun* recollected the accident. "There's a headstone but no body," the reporter notes. Audrey, the widow of pilot Darryl Trottier, had the stone installed in the Fox Creek Cemetery in memory of her husband, but no formal funeral was held. When the local arena was dedicated to the five missing men in 1983, this was considered their memorial service.

On July 23, 2004, shocking news hit the airwaves: a prospector had discovered wreckage above the treeline in the mountainous Monkman Pass near Tumbler Ridge, BC, and reported his findings to the RCMP. Because it matched the flight path of C-GJLJ, and markings on the wrecked aircraft were similar to that of the Cessna 185, excitement mounted that this was indeed the aircraft that had taken the five Fox Creek men to their deaths. Alas, it was found to be from a Cessna 180 (CF-JIV) that had crashed back in 1965. These and similar accidents explain why many pilots steer clear of the Monkman Pass.

Discovery awaits, but as friend Jimmy "Midnight" Anderson notes, "The boys haven't phoned in yet."

THE EVERLASTING HONEYMOON

At 11:13 PM on Monday April 28, 1947, just seventeen minutes before its estimated time of arrival, the captain of Trans Canada Airlines Lockheed Lodestar 18-08A (CF-TDF) called in to the control tower at Sea Island Airport, twelve miles from downtown Vancouver. There was no indication of any problems with the twin-engined aircraft. That was good, because the three-hour Flight 2, eastbound from Vancouver to Lethbridge, and the return Flight 3 were over Canada's highest mountains and encountered some of the strongest winds and worst weather in all the country.

Just the week before, Flight 2 had had to turn around when the wings had iced up over the Kootenays. The Lodestar had diverted to Penticton only to learn that the airport was socked in. "The plane limped back to Vancouver," stated the *Vancouver Sun*. "It had been a remarkable bit of flying by Captain Pike." The stewardess on that flight, Audrey Brandon, admitted that she'd experienced more than a few anxious moments.

Captain W. G. Pike, with First Officer A. A. Stewart, was again flying this route on April 28. Stewardess Brandon had also been scheduled, but was bumped at the last moment and replaced by her friend Helen Saisbury. Flight 2 on April 28 had gone well, and now Flight 3 was coming back to Vancouver. The wind was at fifteen miles per hour out of the east. Although it was raining heavily with a ceiling of 4,600 feet, conditions were not too bad.

"TCA-3 to Vancouver tower," Captain Pike announced on approaching the airport. "By the range at 7,000 on instruments," he said, indicating that he'd passed the radio range, two and one-half miles east of the airfield at Richmond, at seven thousand feet. "Westbound at 11:13." Control tower operator Kenneth St. John noted, "It was just a routine check. The pilot's voice was clear, his tone was normal."[1]

The operator logged the call and waited for the aircraft to make the three-minute-westward circuit, which would take it over the twenty-five-mile-wide Strait of Georgia to drop below clouds and obtain clearance for a landing from the northwest on Runway 11. But the silver-colored airliner with scarlet trim, observed heading out to sea to make its approach turn, never came back. It simply disappeared.

Over the eight-hour search, authorities received 442 calls from people who claimed to have heard or seen something relating to the missing aircraft. Every clue was checked.

Royal Canadian Air Force Search and Rescue planes swept the areas along the Strait of Georgia and also made flights over the heavily forested areas of Vancouver Island up to Nanaimo. By the following day, nineteen aircraft were in the air searching for the missing Lodestar. As well, marine vessels ranging from RCAF rescue launches to fishing boats were alerted by radio-telephone. The US Coast Guard and military parachutists joined in the search, along with navy divers, private aircraft and sea vessels, hunting dogs, mystics, diviners, and readers of tea leaves.

Tips were followed, only to be found inconclusive. An oil slick by Steveston was "a quantity of oil rags." Odd noises remained unidentified. Red glows in the sky noted after 1:00 AM on Tuesday were untraceable.

Airline officials and others in the field could only speculate on what might have gone wrong. Could the airliner's radio have failed, causing it to go off beam, making it impossible to call in?

The rain-swept coastline, buffeted by thirty-three-mile-per-hour winds, was a ghastly host for downed aircraft and their searchers. The ceiling dropped, but still the search continued, focusing on a thirty-square-mile area on Vancouver Island's east coast around Duncan, Chemainus, Nanaimo, and nearby Mount Moriarity, due west of Vancouver. By mid-May, all probable areas had been searched and re-searched, over twelve thousand square miles on the Lower Mainland and Vancouver Island. But the silver airplane with its fourteen occupants was gone.

Immediately following the disappearance of CF–TDF, the newspapers ran the list of crew and passengers. The three crew members were Captain Pike, of Vancouver and Toronto, First Officer Stewart, of Vancouver, and Stewardess Helen M. Saisbury, of New Westminster, BC. Helen, a qualified nurse, as stewardesses were then required to be, was an only child whose aviation adventures had brought both pride and anxiety to her parents.

Revenue passengers hailed from Quebec to BC. One man, Mr. H. Wolf, from London, England, claimed to be a direct descendent of General James Wolfe, who had led the attack to capture Quebec from the French in 1759. Mr. Wolf (whose family had later dropped the "e" from their surname) was making his inaugural trip to Canada, and after making a business call in Vancouver, he planned to visit the Plains of Abraham to pay homage to his ancestor.

Other passengers on the flight included people from across the country. Victor Armand of Vancouver was a superintendent for Famous Players Corporation; a Montreal man with the ominous name of Clarence W. Reaper was on board; Saskatchewan-born Jane Warren and Margaret Hamblin were student nurses at Vancouver General Hospital returning from a vacation; D. Vance and W. Robson were from Manitoba; and Lance Millor, an agent for a manufacturing company, was from Vancouver.

Of the four nonrevenue passengers, two women, Margaret Trerise and Anastasia (Nell) Lesiuk, worked for Trans Canada Airlines. Margaret had grown up in Port Coquitlam, BC, and took her nurse's training at Royal Columbian Hospital. Anastasia had been a stewardess for several months, stationed in Moncton, New Brunswick, and was enjoying seeing the country while being paid to do so. The night of April 28, two of Anastasia's sisters at home in Trail, BC, experienced terrible dreams. Zenida dreamed of a plane crash. Helena dreamed she saw Anastasia's face all black, while she was dressed in white. The dreams frightened the sisters, but they dared not mention them to anyone but each other.

Two other passengers who'd received complimentary tickets were Cecil and Marjorie (Brown) Nugent. Marjorie Brown was employed in the intelligence office of the RCMP, Winnipeg detachment; her new husband, Cecil Nugent, was a former World War II RCAF fighter pilot and currently a TCA employee at the Winnipeg airport. Just married in Winnipeg on Saturday, April 26, they caught TCA flights west on Monday the 28th. The honeymooning couple was making a quick trip out to the west coast to do some sightseeing, then coming back east to Toronto to see Marjorie's sister, Mrs. J. C. (Betty) Smith of Trenton, Ontario.

Supposedly encouraged by the ticket agent, they'd bought flight insurance in Lethbridge before leaving on the final leg of the trip west. There were some anxious moments when they heard that someone had requested one of their seats, but Bill London, who worked with Cecil Nugent, obligingly canceled his flight to allow the groom to complete the trip with his bride. The honeymoon was on.

Marjorie Nugent's sister, Betty Smith, will never forget the day the aircraft disappeared. "I was dressing my two children to go outside and I heard it on the radio. I ran around the house frantically—my stomach went right down to my boots. It was terrible."[2] Betty's recollections of that terrible time remained with her some fifty-seven years later. "I was very close to my sister," Betty Smith wrote to the author in 2004. "I can see us now sitting in

Newlyweds Marjorie and Cecil Nugent. (Courtesy Betty Smith)

This memorial stone marks the final resting place of the TCA Lockheed Lodestar (CF–TDF). (Courtesy Betty Smith)

LOCATION OF AIRCRAFT N 49° 24.34′ W 122° 56.84′

ON APRIL 28, 1947, TRANS-CANADA AIR LINES FLIGHT 3, EN ROUTE FROM LETHBRIDGE, ALBERTA WENT MISSING ON ITS APPROACH TO VANCOUVER, BRITISH COLUMBIA. ITS LAST REPORTED POSITION WAS OVER VANCOUVER AIRPORT AT 23 13 HOURS. THE AIRCRAFT WAS DISCOVERED AND IDENTIFIED IN SEPTEMBER, 1994 IN A REMOTE AREA JUST WEST OF MOUNT ELSAY, ON A SLOPE BEHIND THE RIDGE INDICATED BY THE ARROW.

the kitchen, with our mother teaching us how to knit and embroider. Marge, as we called her, was very talented. She played the piano very well, and was a happy-go-lucky and very nice person. I still miss her very much, and [writing] this sure brings back memories. My sister was a great person and went too early in life. I believe she would have been some kind of a leader."

Marjorie's mother, Mrs. Brown, could not bear the pain of being so far away from where her daughter might be (Marjorie's father had died in 1943), so she moved from Winnipeg to Vancouver, along with her son Jack and his family. Betty recalls, "Mother was never happy, always wondering what happened to the plane. Never a day went by that Mother would always talk about my sister. My husband gave up his job in 1955 and we decided to move to Burnaby, BC, thinking if the aircraft was found we would be close by." Mrs. Brown died in 1964, with the mystery of her daughter's fate still unsolved.

In 1992, an employee of the Greater Vancouver Regional District (GVRD) came across wreckage on a slope just west of Mount Elsay, south of Elsay Lake and due north of Deep Cove, in Mount Seymour Provincial Park. He took some photos and continued on his way. Then, two years later, when hearing about the nearly fifty-year-old disappearance of the big TCA airliner, he reported his findings, stating that "he assumed it was an old wreck that had been marked and done."[3] Employees of the GVRD, guided by their colleague's recollections of the location, went in search of the wreckage.

The Lockheed Lodestar, CF-TDF, was found on September 27, 1994— forty-seven years after it had disappeared. The aircraft had apparently gone into a spin before it crashed and burned. Now it lay at the 3,500-foot elevation of the mountain, surrounded by steep cliffs and thick rain forest. The remains of the passengers and cargo were still present, the journey not yet complete. Along with them, intact and undisturbed over the years, were personal effects including money and jewelry.

Because of the remote and inaccessible location, the investigative team had to be airlifted by helicopter to the site, which forty-seven years of vegetation growth had all but obscured. The aircraft parts that were considered significant and fairly whole were brought back for examination, then donated to the Canadian Museum of Flight and Transportation (formerly in Surrey, now in Langley, BC), as its staff was instrumental in viewing photos of the wreckage and pinpointing its identity.

The airline company (by then Air Canada), sent condolence letters to relatives of the deceased and explained the process being undertaken to retrieve any effects that might be cleaned, restored, and returned to the families. "I received my sister's wedding ring from Air Canada," Betty Smith writes. "How we knew it was her ring was she was the only married lady on the aircraft. I also received a powder compact with her initials on it. Jewelry found was sent to London, England, to be cleaned." Betty states, "To this day I will not get on a plane. If I cannot go by boat, car, or train I will stay home."

The family held a memorial service for Marjorie and Cecil at East Burnaby United Church on October 13, 1993. A formal memorial service was held on Friday, April 28, 1995, in the Seymour Demonstration Forest, with the airline offering transportation and accommodation to any friends or relatives who might wish to attend. A monument was erected in an accessible area near the site.

The investigation was inconclusive, and the reason for the tragedy remains a mystery. Dr. Jack Albrecht of Burnaby, BC, has had a long acquaintance with aviation, and with Mrs. Betty Smith as well. While his sympathy goes to the survivors and relatives of this and other tragedies, he acknowledges a fatal fact: "BC's west coast has a great appetite for aircraft. It swallows them whole."

EPILOGUE

A Libyan fable tells of an eagle stricken with a dart. On the shaft that caused his impending death he spied one of his own feathers. Then came the realization: "With our own feathers, not by other's hands, are we now smitten" (Aeschylus, 525–426 BC).

The philosophic eagle, as well as humans whose lives have ended so abruptly in a fall from the skies, might share this fateful recognition, later immortalized by Lord Byron:

> 'Twas thine own genius gave the final blow,
> And helped to plant the wound that laid these low;
> So the struck eagle, stretched upon the plain,
> No more through rolling clouds to soar again,
> Viewed his own feather on the fatal dart,
> And winged the shaft that quivered in his heart.

—English Bards and Scotch Reviewers, 1809

NOTES AND SOURCES

Quoted material in each of the chapters that is not specifically noted in the text or cited in the following notes is derived from interviews and personal communications between the person quoted and the author.

INTRODUCTION

Notes

1. The late George Greening, from Prince Albert, Saskatchewan, received his air engineer's licence in 1930. He also performed daring feats at North Battleford for the Northern Aero Club as a wing walker on Pheasant and Curtiss Jenny aircraft at country fairs. During World War II, he served with No. 6 Elementary Flying Training School as the assistant chief engineer for both the flying training and air observers schools. In 1943 he qualified for his commercial pilot's licence and began a colorful flying career throughout the North. He survived a cartwheel in a Fairchild 82 on Île-à-la-crosse Lake when his wing tip touched the ice during a whiteout, and he later survived a forced landing in a Waco in the top of a grove of white poplar trees at Cowan Lake. Greening clocked fifteen thousand flying hours during his thirty-two years of bush flying. The poem "Reverie" was submitted by George's brother, Jack Greening, of Christopher Lake, Saskatchewan.

FLIGHT PLANS FOR FREEDOM

Notes

1. The men each faced six charges: breaking out of Headingly Jail, using violence; theft of a $1,000 car, property of the Province of Manitoba; breaking into the farmhouse belonging to Nestor Ewanek and stealing food and clothing worth $90; unlawfully confining Ross Mackenzie of Balmoral, Manitoba; unlawfully confining Heather Jackson of Stonewall, Manitoba; theft of an aircraft valued at $11,000, property of Abe Lowen of Steinbach, Manitoba.

Sources

Handman, Stanley. "He Went On Business Trips – To Rob Banks." *The Albertan*, 19 July 1958.

Redekop, Bill. "The Great Gold Robbery—Flying Bandit Ken Leishman." In *Manitoba's Most Notorious True Crimes, Crimes of the Century*. Winnipeg: Great Plains Publications, 2002.

Robertson, Heather. *The Flying Bandit*. Toronto: James Lorimer & Company, 1981.

STALIN'S FALCONS

Notes

1. Walter Kurilchyk , *Chasing Ghosts* (Capistrano Beach, CA: Aviation History Publishing, 1997), 42.

2. Von Hardesty, "Soviets Blaze Sky Trail Over Top of World," *Air & Space*, Smithsonian (Dec. 1987-Jan. 1988): 48-54.

3. Yuri Kaminsky, "Why Didn't Levanevsky Arrive in America?" trans. Mike J. Hewitt, American Aviation Historical Society, *AAHS Journal* 45, no. 2 (summer 2000): 111–117.

4. *Los Angeles Times,* 17 July 1937.

5. The following are the statistics of the DB-A in metric measurements found in I. E. Negenblya's "Missing on the Ice," chap. XXIII in *Over the Boundless Arctic,* trans. Mike J. Hewitt (Yakutsk, Russia: Publishing House Yakutian Region, 1997):
    • length of aircraft: 24.4 m
    • wingspan: 39.5 m
    • mass of empty aircraft: 15,400 kg
    • flying (loaded) mass: 21,900 kg
    • type, quantity, and power of engines: M-34RN, 4 x 970 hp
    • maximum speed at ground level: 280 k/h
    • maximum speed at an altitude of 4,400 m: 330 k/h
    • practical ceiling: 7,220 m
    • range of flight: 7,000 km

6. *Technica Molodezhy* (or *Technika-Molodezhi* [Technics for Youth]), trans. Mike J. Hewitt (1995), and excerpted from "Why Did N-209 Vanish?"

7. Yuri Kaminsky, "Why Didn't Levanevsky Arrive in America?" 112.

8. Negenblya, *Over the Boundless Arctic.*

9. Zinovy Kanevsky, "Searching for Levanevsky," *Mysteries and Tragedies of the Arctic,* trans. Mike J. Hewitt (1991).

10. Kaminsky, "Why Didn't Levanevsky Arrive in America?" 116.

11. Kaminsky, "Why Didn't Levanevsky Arrive in America?" 116.

12. Eugenie Louise Myles, *Airborne from Edmonton* (Toronto: The Ryerson Press, 1959), 239.

13. Ernst Teodorovich Krenkel, *RAEM Is My Call Sign,* trans. R. Hammond (USSR: Progress Publishers, 1978).

14. General Georgi Baidukov later explained to Alaska pilot Ron Sheardown that Map 34 was in the Canadian Arctic between 70 and 75° North and 85 to 115° West. This covers the Brodeur Peninsula on Baffin Island to Banks Island, including the northern part of Victoria Island. Levanevsky had used Baidukov's maps of the Canadian Arctic for the trip. According to Sheardown, Map 34 was not made available to the search, along with much other information that would have been useful.

15. The American Polar Society, "Levanevsky Down in Arctic Wastes," *The Polar Times,* no. 5 (Oct. 1937): 9, 9B.

16. Kaminsky, "Why Didn't Levanevsky Arrive in America?" 117.

17. Ron Sheardown, e-mail to author, 16 November 2004.

18. "There is also another mystery," notes Mikhail Rebrov, editor of *Red Star.* A reader, Aleksey Ivanovich Vinogradov, informed the editor that one of the radio posts at noon on August 14 picked up the end of a radiogram where the number "83" was repeated three times and contained "AS 9" (an official expression indicating "I temporarily interrupt the transmission"). No one knows where the message came from, for the coordinates were not given when radio communications from N-209 ceased. (Source: An article by A. A. Burykin in which he summarized all references he could find to the wireless messages sent from the N-209. This message was attributed to Nightingale, a wireless operator from Yakutsk, Alexseyev D. A., Novokshonov P. A., *On the Tracks of Mysterious Journeys* [Moscow, 1988].)

    On August 16, transmissions were picked up by radio stations in Irkutsk, Siberia, and Archangel, which Boris Kalashnikov described as "just illegible scraps of working from a radio transmitter on the emergency frequency RL." But surely, N-209 was not flying by this date.

    On August 22, Dalstroy radio station picked up an unclear signal from

"RL." Seven more times such a signal was repeated. RL also called a portable radio transmitter unknown to Moscow, with the call sign DH and WEBB. "Amateur shortwave enthusiasts usually had figure designations," said Aleksey Ivanovich Vinogradov. "What could this mean? Then, on September 24 at 6:32, fragments of radiograms were intercepted in English: 'There is nowhere to land. Good night, pleasant dreams.' What was this? I still wrack my brains but cannot come up with an answer."

19. Lowell Thomas, *Sir Hubert Wilkins, His World of Adventure* (New York, McGraw-Hill Book Company, 1961), 277.

20. Sir Hubert Wilkins, "Our Search for the Lost Aviators," *National Geographic* LXXIV, no. 2 (Aug. 1938): 160.

21. "Wilkins Believes Soviet Fliers Live," *The Polar Times*, no. 5 (Oct. 1937): 10A.

22. Ernst Teodorovich Krenkel, *RAEM is My Call Sign* (USSR: Progress Publishers, 1978)

23. Dr. Homer Flint Kellems, Director, Will Rogers–Wiley Post Memorial Expedition to Alaska,1938, complete report included in Walter Kurilchyk, *Chasing Ghosts* (Capistrano Beach, CA: Aviation History Publishing, 1997), 178–203.

24. Walter Kurilchyk, e-mail to author, 19 January 2003.

25. Bob Isham, e-mail to author, 9 October 2002.

26 Kurilychyk, *Chasing Ghosts,* 146.

27. Kurilchyk, *Chasing Ghosts,* 149, 162–167.

28. Yuri Salnikov, "The N-209 Enigma," *Aerospace Journal* (Nov./Dec. 1996).

29. Kurilchyk, e-mail to author, 9 November 2004.

30 Kurilchyk, e-mail to author, 19 January 2003.

31. Kurilchyk, *Chasing Ghosts,* 171.

32. Kurilchyk, e-mail to author, 15 November 2004.

33. Mike Hewitt, e-mail to author, 26 November 2004.

34. Ron Sheardown has extensive Arctic flying and resource business experience and is the owner of a Russian Antonov An-2 Colt aircraft. In 1997, Sheardown participated in an expedition to the North Pole, and in 1998 was a team member on a transpolar flight that retraced the routes of Carl ben Eielson and Sir George Hubert Wilkins. He has been to Russia about fifty times in the past fifteen years, and has spent more time than any other Westerner with General Baidukov, the second pilot with Chkalov in the ANT-25 flight from Moscow to Vancouver, Washington, and who gave his maps to Levanevsky. Thus, Ron Sheardown has an understandable interest in the Levanevsky mystery and other northern "ghost airships." Fervent fundraising for the search from sources in Russia, the USA, and Canada is ongoing. Sheardown provided the author with information for this story via e-mail correspondence, documents, and a personal meeting.

35. Mikhail Ilves, "Could Levanevsky Change the Route?" trans. Mike J. Hewitt, *Air Transportation*, no. 2 (1993).

36. Robert J. Morrison, *Russia's Shortcut to Fame* (1987), quoted in *Seattle Times*, 23 October 1983.

37. Kaminsky, "Why Didn't Levanevsky Arrive in America?"

38. Georgi Baidukov, *Russian Lindberg—The Life of Valery Chkalov*, trans. Peter Belov, edited and inscribed by Von Hardesty (Washington, DC: Smithsonian Institution Press, 1991).

Sources

American Polar Society. *The Polar Times*, no. 5 (Oct. 1937); no. 13 (Dec. 1941).

Anchorage Daily Times, 13 August 1937.

Baidukov, Georgi. *Russian Lindbergh— The Life of Valery Chkalov*. Translated by Peter Belov. Washington, DC: Smithsonian Institution Press, 1991.

"Bulletin of the Russian Aviation Research Group of Air-Britain." (Dec. 1987): 27.

Fairbanks News-Miner, 22 October 1987; 16 April 1988.

Ferguson, William Paul. *The Snowbird Decades, Western Canada's Pioneer Aviation Companies*. Vancouver: Butterworth & Co., 1979.

Hardesty, Von. "Soviets Blaze Sky Trail Over Top of World." *Air & Space* (Smithsonian) (Dec. 1987 / Jan. 1988).

Jackson, Donald Dale. "Pioneering the Great Circle Route." In *The Explorers*. Alexandria, VA: Time Life Books, 1983.

Kalashnikov, Boris. *The History of Aviation*. Translated by Mike J. Hewitt. 1997.

Kaminsky, Yuri. "Search, or The Story About Embarrassed Times." *Kremlin Flights*.

Kaminsky, Yuri. "Why Didn't Levanevsky Arrive in America?" Translated by Mike J. Hewitt. *AAHS Journal* 45, no. 2 (summer 2000): 111–117.

Kanevsky, Zinovy. "Searching for Levanevsky." In *Mysteries and Tragedies of the Arctic*. 1991.

Krenkel, Ernst Teodorovich. *RAEM Is My Call Sign*. Translated by R. Hammond. USSR: Progress Publishers, 1978.

Kurilchyk, Walter. *Chasing Ghosts*. Capistrano Beach, CA: Aviation History Publishing, 1997.

Los Angeles Times, 11 July 1937; 17 July 1937.

Morrison, Robert J. *Russia's Shortcut to Fame*. 1987.

Myles, Eugenie Louise. *Airborne from Edmonton*. Toronto: The Ryerson Press, 1959.

Negenblya, I. E. "Missing on the Ice." Chap. XXIII in *Over the Boundless Arctic*. Translated by Mike J. Hewitt. Yakutsk, Russia: Publishing House Yakutian Region, 1997.

Nome Daily Nugget, 13 August 1937.

Salnikov, Yuri. "The N-209 Enigma." *Aerospace Journal* (Nov./Dec. 1996).

Stefansson, Vilhjalmur. *Unsolved Mysteries of the Arctic*. New York: MacMillan, 1939. Reprint, New York: Collier Books, 1967.

Thomas, Lowell. *Sir Hubert Wilkins, His World of Adventure*. New York: McGraw-Hill, 1961.

Whitehorse Star, 2 August 1988.

Wilkins, Sir Hubert. "Our Search for the Lost Aviators." *National Geographic Magazine* LXXIV, no. 2 (Aug. 1938).

Yeletsky, Viktor. "The Search for Levanevsky's Aircraft." *Air Transport*, 3 January 1999.

THE NORTH RECLAIMS ITS OWN

GAVIN EDKINS—A FAMILY'S HOPE, A COMMUNITY'S VIGIL

Notes

The parallel story appearing in this chapter, set in italics throughout, was written by Marion Kuziemsky and was sent to the Edkins family for the record. Permission to use Kuziemsky's narrative was given by both Stan Edkins and Marion Kuziemsky.

1. In a report dated May 19, 1996, Captain G. L. Illchuk summarized recommendations re: Case T96-A0250, including the necessity for detailed flight plans/notices to be filed by all pilots. Also, the use of transponders during VFR flying were stressed for all pilots. If the aircraft transponder had been activated in this case, it would have provided an accurate LKP (likely known position) down-track and a planned routing.

   But local pilots who read the report thought it contained more theory than practice. One of them states: "I don't think the military understands how transponders and VHF radios work on a light aircraft which is 'scud running' 500 feet above ground, down a valley, diverting around clouds or rain. If you have to go low under weather, the VHF radio doesn't have 50 miles' range to report changes. Further, no station at Fort Smith has radar capable of seeing transponder signals on this type of flight."

2. Julie Kosztinka, "Search for Missing Pilot Continues," *Slave River Journal* 20, no. 2 (28 May 1996).

3. In actuality, Stan Edkins is six feet tall and slim at 150 pounds. He has never worn camouflage outfits, or a hat or cap over his grey hair.

4. James Carroll, "Private Search for Lost Pilot Continues," *Slave River Journal* 20, no. 6 (25 June 1996).

5. John Payne was concerned that the first important days were squandered as civilian search planes sat idle due to poor communication and direction from the military; as a result, bad feelings developed and many hours were lost. Aircraft had come from Calgary, Drayton Valley, and the Edmonton area—all of whom had to pay their own expenses. With a waiting period of five to six days before available aircraft were finally put to use, these expenses became

onerous.(Source: A report submitted by John Payne, Fort Smith, NWT, to Directors of Transport Canada, 26 June 1996, from the archives of Stan Edkins.)

6. Anthony Kovats, "Father Still Missing," *Wetaskiwin Times Advertiser*, 22 July 1996: p. 2.

7. In 1998, Stan Edkins received a call from the fisherman who had hooked the hair in Winifred Lake, asking why nothing had been done about his find. Following his conversation with the fisherman, Stan wrote a letter, dated 9 June 1998, to his Member of Parliament, the Honourable Ethel Blondin-Andrew, wherein he laid out the causes of his distress, including his anger at not being notified by the RCMP about the hair. He concluded:

   "As you know, Ethel, in my more than 30 years as a bush pilot in the north, I have been involved in numerous Medi-Vacs and have participated in many searches where there was civilian as well as RCMP participation. Never have I been treated in such a condescending manner. I am a grieving parent; I am not an emotional idiot. The finding of human hair in the lake indicates that there is a body there. If it is not Gavin, then it is someone else. I am asking you if there is anything you can do to ensure that the RCMP conduct a thorough search of Winifred Lake this year."

   Stan Edkins says he did not receive a reply to his entreaty and to his knowledge nothing was done to further his case from the office of the member of parliament. The RCMP organized divers to search the lake that summer. Nothing was found.

   In the late fall of 2004, Stan's daughter Lorell contacted an RCMP officer who was stationed in Fort Smith at the time of Gavin's disappearance, and who is still active in the force but is now stationed elsewhere. She asked for his assistance in finding out if the hair sample is still available, and if new tests could be arranged. With new technology, not available previously, perhaps the hair sample retrieved from Winifred Lake could reveal DNA information vital to the case. (This information was received during a telephone conversation between Stan Edkins and the author on January 2, 2005.)

Sources

Search and Rescue Operations Report, National Defence, (3385-1 SAR EDKINS), Captain Greg Illchuk, 435 (T&R) Squadron, Trenton Search and Rescue Region, Case #0250, 19 May, 1996.

THE GLITTER OF GOLD

Notes

1. Dean Rossworn's statements, *Rossworn v. Wardair Limited*, Notice to Discover Documents, 15 December 1960; Statement of Claim 16 (1/9/60), Affidavit (21/3/61), Order (23/3/61), Notice of Discontinuance (31/5/61), Statement of Defence (16/9/60), and Examination for Discovery (22/3/61).

2. Jim McAvoy, interview and telephone conversations with author, also written and video documents pertaining to the discovery of the crashed Fairchild CF-MAK, made available by Jim McAvoy.

Sources

Department of Transport, Air Services, Civil Aviation Branch (Aircraft Disappearance Report #2297 re: Fairchild 82-D, CF-MAK [ski plane]), 9 June 1964.

"Killed on Impact." *Edmonton Sun*, 9 August 2003, p. 20.

Loome, Jeremy. "39-Year Riddle Solved." *Edmonton Sun*, 6 August 2003, p. 3.

Mullen, Conal. "Plane Hit Hard, Burned in 1964 Crash." *Edmonton Journal*, 15 August 2003.

Sinnema, Jodie. "39 Years Later, Lost Plane of

Legendary Bush Pilot Finally Found." *Edmonton Journal*, 6 August 2003, sec. A, p. 1.

Sinnema, Jodie. "Discovery May End Arctic Mystery." *National Post*, 7 August 2003, sec. A, p. 3.

Sinnema, Jodie. "Pilot Liked To Do Things a Little Differently," *Edmonton Journal*, 8 August 2003, sec. A, p. 2.

"Shock, Excitement, Relief Over Discovery of Missing Plane." *Edmonton Journal*, 9 August 2003, sec. A, p. 2.

Vanderklippe, Nathan, CanWest News Service, Yellowknife NWT, "Funeral for Legendary Pilot 49 [sic] Years After He Died." *Edmonton Journal*, 15 August 2003.

THE HAUNTED FLIGHT PATH

Notes

1. Gordon Root, "Search for Missing Transport Plane Continues," *Vancouver Daily Province*, 23 December 1942, p. 10.

2. "Light on Mountain and Smoke Signals Shift Plane Hunt," *Vancouver Daily Province*, 22 December 1942, pp. 1, 2.

3. Stanley Burke, Jr., "Grave-Faced Pilots Press Hunt for Their Missing Comrades," *Vancouver Daily Province*, 23 December 1942.

4. Burke, "Grave-Faced Pilots."

5. James Fairley, "Reporter Flies Over Cloud-Shrouded Scene," *Vancouver Daily Province*, 9 June 1945, pp. 1, 2.

6. Philip Smith, *It Seems Like Only Yesterday: Air Canada, the first 50 years* (Toronto: McClelland & Stewart, 1986), 246.

7. Smith, *It Seems Like Only Yesterday*, 246.

8. Ralph Langemann, interview with author, 7 January 2002 and other dates up to 19 December 2004. Langemann's biography can be found in Shirlee Smith Matheson's "Tales from the Log Book," in *Flying the Frontiers Vol. I* (Calgary: Fifth House Publishers, 1994), 150-173.

9. *Vancouver Sun*, 8 June 2002.

Sources

Accident Report, Department of Transport, Canada, Serial #56.16: re: Trans-Canada Airlines, DC-4M2, CF-TFD (Dec. 9/56).

Bain, D. M. *Canadian Pacific Air Lines - Its Story & Aircraft.* Calgary: Kishorn Publications, 1987.

Keith, Ronald A. *Bush Pilot With a Briefcase.* Toronto: Doubleday Canada Limited, 1972.

Pigott, Peter. *National Treasure, The History of Trans Canada Airlines.* Madeira Park, BC: Harbour Publishing, 2001.

Smith, Philip. *It Seems Like Only Yesterday, Air Canada, the First 50 Years.* Toronto: McClelland & Stewart, 1986.

Vincent, C. R. "Consolidated Liberator & Boeing Fortress." In *Canada's Wings*, vol. 2. Stittsville, ON: Canada's Wings, 1975.

later, it was brought to Edmonton and restored by volunteers at the Alberta Aviation Museum. The restoration story is included in Shirlee Smith Matheson's "The Merger of the Masters," in *Flying the Frontiers Vol. III* (Calgary: Detselig Ent. Ltd.), 253-54.

3. "Search for Missing Johnnie Bourassa Still Continuing–RCAF Sweeps Over Wide Area," *Yellowknife Blade* vol. 8, no. 12, 4 June 1951, p. 1.

4. "Search Starts As Note Shows Bourassa Unhurt," *Edmonton Journal*, September 1951.

5. "Where Is John Bourassa?" *The Machinery Record* 1, no. 3, 17 September 1951, p. 1.

Sources

Air Accident Report, Department of Transport, and related correspondence, various dates, 23 May 1951 to 6 December 1966.

Peace River Remembers. Peace River, AB: Sir Alexander Mackenzie Historical Society, 1984.

Western Canadian Aviation Museum. *Aviation Review* vol.18, no. 3 (September 1992): 11.

WHERE'S JOHNNY?

Notes

1. In addition to the DFC and Bar, Bourassa also received the 1939–45 Star, Aircrew Europe Star, France and Germany Star, Pathfinder Badge, and Canadian Volunteer Service Medal and Clasp. He was honorably released from service on July 4, 1946.

2. The Fairchild 71C, CF-ATZ was resurrected from its watery grave and winched onto the shoreline. In 1980, thirty-one years

THE SAGLEK SAGA

Notes

1. The entry paraphrasing Llewellyn's novel appears in a transcript of the diary found in Canadian archives, with the indication it was made in quite different handwriting from the other entries. Oddly, the official American transcript housed at Maxwell Airforce Base in Alabama does not include this passage. As well, a number of date entries differ in the typewritten transcripts found in the Canadian and American archives.

Welsh writer Richard Llewellyn's novel, *How Green Was My Valley*, was first published in 1939 and became an instant hit. During World War II, publishers donated thousands of copies of the book for use by military personnel wherever they were stationed. The poetic excerpt, on which the writer based the piece found in Hodge's diary, can be found on page 216 of the 1968 New American Library edition.

2. Newfoundland, then part of Great Britain, was patrolled by the Newfoundland Rangers, which were later incorporated into the RCMP when Newfoundland became a province of Canada in 1949.

3. Among the listed items belonging to 1st Lt. Grover C. Hodge Jr. were a will, a power of attorney, $71 in cash, a Bible, wings, a Gruen wristwatch, and a gold ring. (No mention was made of a camera, film, or diary in his report; the cornet that Hodge played while being held over in Greenland was not his own.)

John Rose notes, in e-mail correspondence on 21 August 2003, that prior to deployment for overseas duty, crew members usually left envelopes with their commands, as well as personal notes to relatives. The aircraft commander was responsible for control and censoring of all such materials.

4. Norman Vaughan is the last surviving member of Admiral Byrd's 1927 expedition to the South Pole. He will be one hundred years of age on 19 December 2005, and plans to again climb Mount Vaughan on that date.

5. William P. Schlitz, "USAF's Survival School: Giving Downed Aircrews a Fighting Chance," *Airforce*, January 1981.

6. In 1970, Air Defence Command Saglek Air Station (Labrador) was established at Saglek, but the site was closed in 1976 and transferred to the Province of Newfoundland. A lower camp was taken over by Aquataine, an exploration company doing offshore oil drilling, and the surveillance station was manned by USAF officers and enlisted technicians. In the mid-1980s, Frontec Logistics Corporation took over the North Warning Radar Site at Saglek, replacing the upper site with new equipment and continuing to operate an active site supplying logistical support to the North Warning (Radar) System.

7. See http://www.lswilson.ca.

8. Some comparisons might be made between the fate of *Time's a' Wastin'* and *Lady Be Good*, a US Army Air Force FB-24D Liberator bomber (376th Bomb Group) that failed to return to its base after attacking Naples from Soluch, Libya, on April 4, 1943.

Sixteen years later, in 1959, the aircraft and bodies of the nine crew members were discovered in the Sahara desert. Two of the crew (2nd Lieutenant Robert F. Toner, co-pilot, and Technical Sergeant Harold S. Ripslinger, engineer) had kept diaries.

Life magazine printed the story on March 7, 1960, along with a two-page aerial photograph of the downed aircraft and a reproduction of the diary's four opened pages in the author's handwriting. A photo of T. Sgt. Ripslinger's handwritten entries can be viewed in Dennis E. McClendon's book, *The Lady Be Good, Mystery Bomber of World War II* (California: Aero Publishers Inc., 1962), with epilogue, 1982. This tragedy also gave rise to improved military survival schools—in this case, for desert climates.

9. Alma H. Rose and John C. Rose, e-mail to author, 21 August 2003.

Sources

Canadian Broadcasting Corporation. St. John's, Newfoundland. *Land and Sea* television program, "The Bonds of Earth" (from the poem, *High Flight*) parts 1 and 2. Bob Wakeham, Producer.

Cavalier for Men. "Dead Man's Log" (Jan. 1954).

Dmitri, Ivan. *Flight to Everywhere—The Picture Journey of Ivan Dmitri — Over 32,000 Miles of Air Transport Command Routes through Jungle, Desert and Arctic.* New York: McGraw-Hill Book Company, Inc., 1944.

Douglas, Dr. W. A. B. (Alex). "The Nazi Weather Station in Labrador." *Canadian Geographic*, Dec. 1981–Jan. 1982.

Lindbergh, Charles A., and Anne Morrow. "Flying Around the North Atlantic." *National Geographic*, September 1934.

McClendon, Dennis E. *The Lady Be Good, Mystery Bomber of World War II.* California: Aero Publishers Inc., 1962. Reprint with epilogue, 1982.

Oyster, Harold E. and Esther M., eds. "Diary of One Now Dead." In *The 319th in Action, the Official History of 319th Bombardment Group in Northwest Africa, Mediterranean and Pacific*, 1976.

Pennington, Robert. "Lost Marauder." *Warbirds International* (fall 1987). Reprinted in *Ghost Warriors of World War II*, Challenge Airwar Special, vol. 1, no. 1 (1994).

Schlitz, William P. "USAF's Survival School: Giving Downed Aircrews a Fighting Chance." *Airforce*, January 1981.

Simonsen, Clarence, and Jeffrey L. Ethell. *Aircraft Nose Art From World War I to Today.* St. Paul, MN: Motorbooks International, 2003.

UNDERWATER PHANTOMS: SECRETS OF THE DEEP

INTRODUCTION

Notes

1. See "Whose Planes are They, Anyway?" http://members.tripod.com/~manchuri-anhitchcock/asnov98.html.

THE FATEFUL FLIGHT 401

Notes

1. See http://www.123gold.com/flight401/letters1.htm.

2. According to John Blatherwick's *A History of Airlines in Canada* (Toronto: Unitrade Associates, 1989), fleet numbers (F/N) 501 and 503 were purchased by Air Canada in 1973, and were involved with leases to Air Transat and Eastern Air Lines until 1981. The crash in the Everglades occurred in December 1972, so it is possible that parts from Eastern's L-1011 *could* have been used in these two aircraft.

3. John G. Fuller, *The Ghost of Flight 401* (New York: Berkley Publishing Corporation, 1976), 187.

Sources

Australian Aviation. *Air Disasters* 1. (1995): 98–111.

Fuller, John G. *The Ghost of Flight 401.* New York: Berkley Publishing Corporation, 1976.

THE PHANTOM LOCKHEED HUDSON

Notes

1. Lockheed produced twenty-five hundred Hudsons during wartime for overseas military use, and approximately sixty aircraft were stationed at the operational training

unit at Greenwood, Nova Scotia. Antisubmarine patrols were formed in the Bay of Fundy and off the south coast of Nova Scotia to seek out German U-boats plying the waters off Canada's east coast. In total, the Hudsons recorded at least twenty-five successes against U-boats when flown by the RCAF, RAF, RAAF, and RNZAF in every maritime theater of war ranging from the Mediterranean, South Pacific, and Indian Oceans, to the North Atlantic and south to the east coast of the United States. The Hudson's success and reliability for bringing back crews earned it the fond nickname, "Old Boomerang."

Sources

Foot, Richard. "Finding a Hudson Would Be Like Winding Back the Hands of Time." *National Post*, Jan./Feb. 2002, sec. B, p. 6.

Robinson, Jennifer. "Museum Renews Efforts to Raise Bomber." The Canadian Press, 6 January 2002.

GHOST HUNTERS

Notes

1. Colonel Joseph Bradley McManus, USAAC (Ret'd), is quoted in James P. Busha's "Ice Capades–Arctic Adventures in a P-38F," *Flight Journal* 9, no. 1 (Feb. 2004).

Sources

Busha, James P. "Ice Capades–Arctic Adventures in a P-38F." *Flight Journal* 9, no. 1 (Feb. 2004).

Hayes, David. *The Lost Squadron*. Toronto: Madison Press Books, 1994.

Hinton, Steve. "Frozen in Time: Flying the P-38F 'Glacier Girl.'" *Flight Journal* 9, no. 1 (Feb. 2004).

BASHIN' BILL BARILKO

Notes

1. Douglas Hunter, *Open Ice, The Tim Horton Story* (Toronto: Penguin Books Canada Ltd., 1994), 5. Quotes used with author's permission.

2. Hunter, *Open Ice, The Tim Horton Story*, 89.

3. The Toronto Maple Leafs won the Stanley Cup in 1917–18, 1921–22, 1931–32, 1941–42, 1944–45, 1946–47, 1947–48, 1948–49, and 1950–51. Their next win was in 1961–62.

4. Hunter, *Open Ice, The Tim Horton Story*, 160.

5. Hunter, *Open Ice, The Tim Horton Story*, 174–75.

6. Hunter, *Open Ice, The Tim Horton Story*, 439.

7. Brian McFarlane, *Brian McFarlane's World of Hockey* (Toronto: Stoddart Publishing Co., 2000), 223–26. Quotes used with author's permission.

8. Department of Transport, Air Services, Accident Report Serial #306 re: CF-FXT; and Larry Milberry, *Austin Airlines, Canada's Oldest Airline* (Toronto: CANAV Books, 1985), 70.

9. John Melady, *Overtime, Overdue: The Bill Barilko Story* (Trenton, ON: City Print, Publishing Division of Almey Press Ltd., 1988), 118.

10. Melady, *Overtime, Overdue*, 133.

11. *Toronto Sun*, 21 April 2001.

12. The term "fifty mission cap" originated in World War II. Pilots who had flown more than fifty missions beat and "worked in" their usually perfect caps to indicate they were well-used and the wearers were experienced pilots. A possible connection

between Gord Downie's song and Barilko's hockey card being worked in like a "fifty mission cap" might be the large number of missions, or flights, spent in searching for Barilko's missing aircraft. Or, like Andrew Hunter's take on the story, it could be fiction and myth integrated to create an even more intriguing tale, and a memorable song.

13. Hunter, *Open Ice, The Tim Horton Story*, 410.

14. *Toronto Sun*, 21 April 2001.

Sources

"Gleaming Metal Helped Searchers Find Wreckage." *Toronto Star*, 8 June 1962.

Hunter, Andrew. *Up North, A Northern Ontario Tragedy*. Book and exhibition presented at the Tom Thomson Memorial Art Gallery, Owen Sound, ON, 1997.

Hunter, Douglas. *Open Ice, The Tim Horton Story*. Toronto: Penguin Books Canada Ltd., 1994.

McFarlane, Brian. *Brian McFarlane's World of Hockey*. Toronto: Stoddart Publishing Co., 2000.

Melady, John. *Overtime, Overdue: The Bill Barilko Story*. Trenton, ON: City Print, Publishing Division of Almey Press Ltd., 1988.

Milberry, Larry. *Austin Airlines, Canada's Oldest Airline*. Toronto: CANAV Books, 1985.

O'Connor, Joe. "Love Lost in the Wilderness." *National Post*, 11 December 2004, sec. A, pp. 1, 9.

BRITISH COLUMBIA'S MISSING IN ACTION FILES

DEATH FROM ABOVE

Notes

1. Dirk Septer, "Broken Arrow: Many Questions Remain," *West Coast Aviator* (July/Aug. 1998):15.

2. Jeff Nagel, "Lingering Nuclear Mystery Remains–Experts Fly to Crash Site," *Terrace Standard*, 2 October 2002, sec. A, p. 3.

3. Jeff Nagel, "Special to the News," *Interior News*, 16 October 2002, sec. A, p. 3.

4. *Terrace Standard*, 2 October 2002. Nagel, "Special to the News," sec. A, p. 3.

5. Jeff Nagel, "Special to the News," *Interior News*, 16 October 2002, sec. A, p. 3.

6. "Rangers' Crash Site Mission a Blast," *Northern Sentinel*, 6 November 2002, p. 3.

7. Jeff Nagel, "Our Atomic Secrets," *Terrace Standard*, 10 March 2004, p. B1.

Sources

7th Bombardment Wing Operations. Carsewell AFB, 1949-51 "Second Wing B-36 Accident." http://www.7bwb-36assn.org/b36genhistpg2.html.

Baker, David. "Database–Convair B-36 Peacemaker." *Aeroplane*, December 2000, 23-37.

Davidge, Doug. "Environmental Impact Study of Crash Site of USAF Bomber." Environment Canada.

Nagel, Jeff. "Crash Site Contained More Explosives Than Expected." *Interior News*, 16 October 2002, sec. A, p. 10.

Nagel, Jeff. "Lingering Nuclear Mystery Remains–Experts Fly to Crash Site." *Interior News*, 9 October 2002, sec. A, p. 3.

Nagel, Jeff. "Our Atomic Secret." *Terrace Standard,* 10 March 2004, sec. B, p. 1.

Ricketts, Bruce. "Broken Arrow, A Lost Nuclear Weapon in Canada." http://www.mysteriesofcanada.com/BC/broken_arrow.htm.

Septer, Dirk. "Broken Arrow." *BC Aviator* (Oct./Nov. 1993).

Septer, Dirk. "Broken Arrow: Many Questions Remain." *West Coast Aviator* (July/Aug. 1998): 15.

The Haunted Island

Notes

1. Jane Gaffin, *Edward Hadgkiss Missing in Life* (Whitehorse, Yukon: Word Pro, 1989), 119.

2. Tessa Derksen, "Missing in Life," *Aviator Magazine*, October 2000, 42–44.

Sources

Derksen, Tessa. "Missing in Life." *Aviator Magazine*, October 2000, 42–44.

Gaffin, Jane. *Edward Hadgkiss Missing in Life.* Whitehorse, Yukon: Word Pro, 1989. Quotes used with author's permission. Word Pro, Box 31744, Whitehorse, Yukon, Y1A 6L3 (toll free: 1-800-661-0508)

The Lost Boys

Sources

Barrett, Tom. "Is God Taking Care of Daddy?" *Edmonton Journal*, 13 May 1982, sec. A, p. 1.

Olsen, Tom. "Plane Disappeared in '82 Without a Trace–Vanished into Thin Air." *Edmonton Sun*, 17 November 1991, p. 32.

Tetley, Deborah, and Dwayne Erickson. "Old Plane Wreck May Contain Alberta Men." *Calgary Herald,* 23 July 2004, sec. A, p. 5.

Tetley, Deborah. "Plane Wreck Misidentified as 1982 Crash." *Vancouver Sun*, 24 July 2004, sec. B, p. 11.

Tetley, Deborah. "Plane Wreck Still Unsolved After Bizarre ID Mistake." *Calgary Herald*, 24 July 2004, sec. A, p. 3.

The Everlasting Honeymoon

Notes

1. "T. C. A. Plane Lost at Coast; Oil Slick Spotted in Straits," *Calgary Herald*, 29 April 1947, p. 1.

2. Anna Marie D'Angelo, "Fatal Air Crash Site Discovered–Downed Plane Found on Mt. Seymour," *Vancouver Sun*, 30 September 1994.

3. D'Angelo, "Fatal Air Crash Site Discovered–Downed Plane Found on Mt. Seymour."

Sources

"4 Clues Found in Airliner Search." *Vancouver Sun*, 29 April 1947.

Bell, Stewart, and Al Sheehan. "Wreckage from Crash in '47 Believed Found in Seymour Forest." *Vancouver Sun*, sec. A, p. 1.

Kent, A. B. "Chart Pins Record Lost Planes for Past 36 Years." *Victoria Times*, 21 October 1976, p. 18.

McMartin, Pete. "Stewardess Bumped From Fateful Flight Recalls Friend Who Never Came Back." *Vancouver Sun.*

"'Mystical Powers' Offered to Locate Lost T. C. A. Plane." *Vancouver Daily Province*, 15 May 1947.

"Still No Clue to Lost Plane." *Calgary Herald,* 30 April 1947.

About Fifth House

Fifth House Publishers, a Fitzhenry & Whiteside company, is a proudly western-Canadian press. Our publishing specialty is non-fiction as we believe that every community must possess a positive understanding of its worth and place if it is to remain vital and progressive. Fifth House is committed to "bringing the West to the rest" by publishing approximately twenty books a year about the land and people who make this region unique. Our books are selected for their quality, saleability, and contribution to the understanding of western-Canadian (and Canadian) history, culture, and environment.

Look for the following Fifth House titles at your local bookstore:

Flying the Frontiers: A Half-million Hours of Aviation Adventure,
Shirlee Smith Matheson

Flying Under Fire: Canadian Fliers Recall the Second World War, selected and edited by William J. Wheeler of the Canadian Aviation Historical Society

Flying Under Fire Volume Two: More Aviation Tales from the Second World War, selected and edited by William J. Wheeler of the Canadian Aviation Historical Society

Pilots of the Purple Twilight: The Story of Canada's Early Bush Flyers,
Philip H. Godsell

Skippers of the Sky: The Early Years of Bush Flying, selected and edited by William J. Wheeler of the Canadian Aviation Historical Society